Wikipedia and the Politics of Openness

Wikipedia and the Politics of Openness

NATHANIEL TKACZ

The University of Chicago Press
Chicago and London

Nathaniel Tkacz is assistant professor in the Centre for Interdisciplinary Methodologies at the University of Warwick and coeditor of *Critical Point of View: A Wikipedia Reader*.

The University of Chicago Press, Chicago 60637
The University of Chicago Press, Ltd., London
© 2015 by Nathaniel Tkacz
All rights reserved. Published 2015.
Printed in the United States of America

24 23 22 21 20 19 18 17 16 15   1 2 3 4 5

ISBN-13: 978-0-226-19227-7 (cloth)
ISBN-13: 978-0-226-19230-7 (paper)
ISBN-13: 978-0-226-19244-4 (e-book)
DOI: 10.7208/chicago/9780226192444.001.0001

Library of Congress Cataloging-in-Publication Data

Tkacz, Nathaniel, author.
 Wikipedia and the politics of openness / Nathaniel Tkacz.
      pages cm
 Includes bibliographical references and index.
   ISBN 978-0-226-19227-7 (cloth : alkaline paper)—ISBN 978-0-226-19230-7 (paperback : alkaline paper)—ISBN 978-0-226-19244-4 (e-book)   1. Wikipedia—Social aspects. 2. Authorship—Collaboration—Social aspects. 3. User-generated content—Social aspects. 4. Communities of practice.   I. Title.
 AE100.T53 2015
 030—dc23
                                                                 2014010459

♾ This paper meets the requirements of ANSI/NISO Z39.48–1992 (Permanence of Paper).

For Ernest

Contents

Acknowledgments ix
Introduction 1

1 **Open Politics** 14
2 **Sorting Collaboration Out** 42
3 **The Governance of Forceful Statements:
From Ad-Hocracy to** *Ex Corpore* 88
4 **Organizational Exit and the Regime of Computation** 126
5 **Controversy in Action** 150

Conclusion: **The Neoliberal Tinge** 177

Appendix A: **Archival Statements from the Depictions
of Muhammad Debate** 183

Appendix B: **Selections from the Mediation Archives** 193

References 199
Index 209

Acknowledgments

Research for this book began at the University of Melbourne in 2006. Five years of lively discussions with friends and colleagues at Melbourne helped shape the project and its intellectual flavor. Thank you to Michael Dieter, Tom Apperley, Bjorn Nansen, Nicole Heber, and to everyone else who gathered at Tsubu on Fridays (Graham, Janine, Jacob, Rachael, Darshana, Dale, Marcos, Luke VR, Luke H, Robbie, Justin, Nick, Emmett, and others). I am especially grateful to John Frow and Sean Cubitt for their close readings and feedback at various stages.

In 2009 I met Geert Lovink for the first time and together with Nishant Shah and Johanna Niesyto we formed the Critical Point of View (CPOV) Wikipedia Research Network. Since then, Geert has become a regular collaborator and friend, and continues to be an inspiration. The CPOV network offered a space to think and discuss all things Wikipedia where neither evangelical nor conspiratorial tendencies prevailed. Many of the themes discussed at CPOV events are present in the pages that follow.

In 2012 I joined the Centre for Interdisciplinary Methodologies at the University of Warwick, where this book was completed. Thank you to my colleagues Celia Lury, Emma Uprichard, Olga Goriunova, and Will Davies for conversations that helped me finish the book and for providing a stimulating place to work within the hazy English Midlands.

Ned Rossiter and Alan Liu also read full versions of the manuscript and made many helpful comments, as did two anonymous reviewers for the University of Chicago Press. My gratitude goes to Alan Thomas and the staff at the Press for showing interest in the manuscript and seeing it through to publication. Early versions of bits and pieces of chapters 1, 2, and 3 were published in *Ephemera, Fibreculture, Digithum* and *Platform*. I thank the editors and reviewers of these journals for their feedback.

Finally, this book would not be possible without the support of family and friends. Thanks to my wonderful partner, Mia, and our little son, Ernest.

Introduction

On January 21, 2009, the day after his inauguration, President Barack Obama signed a "Memorandum on Transparency and Open Government." The first few lines of the memorandum state:

> My Administration is committed to creating an unprecedented level of openness in Government. We will work together to ensure public trust and establish a system of transparency, public participation, and collaboration. Openness will strengthen our democracy and promote efficiency and effectiveness in Government.[1]

At the time, the occasion was largely overshadowed by the series of monumental announcements President Obama would make in the first weeks of his presidency: the closure of the detention camp at Guantanamo Bay; the drawing up of an exit plan for military operations in Iraq; the revision of equal-pay laws; expanded healthcare for children; the revision of policies relating to embryonic stem-cell research; and the appointment of the first ever Hispanic Supreme Court Justice (who also happened to be female). Faced with the first wave of the fallout from the global financial crises—one that still shows few signs of clearing—Obama also announced a massive economic stimulus package and made heavy interventions in the automotive industry. While the details of these announcements were often complicated and their resulting implementations messy, compromised, and ultimately underwhelming, they nevertheless all responded to what the new government saw as specific political problems: the health of children, the costs of war, the future of science,

---

1. B. Obama, "Transparency and Open Government" (2009), accessed April 20, 2010, http://www.whitehouse.gov/the_press_office/TransparencyandOpenGovernment/.

the equality of citizens in work and jurisprudence, the economic future of the nation, and the security of its industries. If all of these announcements had a certain specificity to them and operated within well-established political problematics, the memorandum, which also marked the beginning of a more general "Open Government Directive" and later "Open Government Initiative," was of a somewhat different nature. Its focus was directed inward: not on what Obama's administration should or would do, but on what it should be. It was a statement about politics and governance in general.

While the old political stalwart, democracy, still featured in the opening lines of the memorandum, and while the market principles of efficiency and effectiveness were also still present, this new political mode of existence was clearly not primarily about these things. The practice of government was now about creating openness, and openness would in turn keep the older and evidently weak notions like democracy strong. The future of democracy and the proper functioning of government now relied on their being open. The way to achieve this new openness, as detailed in the rest of the memorandum, was through a set of closely related notions: transparency, participation, and collaboration. Each paragraph of the memorandum begins with a strong declaration, written in bold font, about the new mode of being of government:

> Government should be transparent . . .
> Government should be participatory . . .
> Government should be collaborative . . . [2]

What is immediately striking about these terms is that all have previously been deployed to define developments in network and software cultures. It's as if Obama's description of Open Government was channeling Tim O'Reilly's definition of Web 2.0 a few years earlier—and they both sounded a lot like Eric Raymond commenting on open source software cultures of the 1990s. In fact, this connection would be made explicit in the book *Open Government: Collaboration, Transparency, and Participation in Practice* (Lathrop and Ruma 2010) published by O'Reilly Media. O'Reilly's own contribution to the book draws directly on Raymond's distinction between cathedral- and bazaar-style software development, but reapplies it to the world of politics. After noting, "In the technology world, the equivalent of a thriving bazaar is a successful platform," for example, O'Reilly posits the following as the new question for governance: "How does government become an open platform that allows people inside and outside government to innovate?" (O'Reilly 2010, 13). For O'Reilly, the open platform bazaar is the new model for governance.

2. Ibid.

Of course, dreams of technological solutions to political problems are hardly new. But even though Obama directed his chief technology officer to coordinate the Open Government Directive, and even though he wrote of "harness[ing] new technologies" and "innovative tools"[3] in the memorandum, it would be wrong to couch the Open Government Initiative in terms of a technological fix. Rather, a whole set of practices and ideas, an entire Weltanschauung, that had already flowed through network cultures, had now entered the discourses of institutional politics. And it would equally be a mistake to frame the arrival of openness as specific to the official voice of power—that is, as pure ideology. Rather, openness had become part of the conditions of possibility of all politics. It surely meant different things to different political actors, but it meant something to everyone. Openness had become a key coordinate for steering the political–cybernetic connotations implied. Toward the end of 2011, for example, a banner appeared at the top of the information-leaking activist site WikiLeaks. The banner featured the site's notorious spokesperson, Julian Assange, asking for donations. The text next to Assange read: "Help WikiLeaks Keep Governments Open." And when WikiLeaks defectors Daniel Domscheit-Berg and Herbert Snorrason started a competing info-leaking site, they called it "OpenLeaks."

This book offers a consideration of political developments that operate under the notion of openness. It is not a book on how to be open, nor does it seek to validate some things as open while finding others lacking—a task that would only become possible by first establishing a comprehensive criterion from which to adjudicate the truth of being open. It does, however, unfold in a political and intellectual climate where the question of how to be open has become a central one, and where the truth of openness has become hotly contested. Rather, this book offers what might be called a politics in the face of openness—a politics in spite of openness. The precise reasons for this approach form a major theme of the book, but it's worth sketching a few of them by way of introducing how I came to be interested in openness in the first place.

Like many other people, in the period between the dotcom burst and the peak of Web 2.0 hype in the mid-2000s, I became interested in Wikipedia. To borrow the subtitle of Andrew Lih's book on the subject, I too was interested in "how a bunch of nobodies created the world's greatest encyclopedia" (2009). The project was obviously of a different breed to the startups that had gone under in the burst, but what was interesting was that it seemed equally different to ones that had survived the burst or were emerging in its

3. Ibid.

aftermath: Amazon, eBay, Google, Myspace, Facebook, YouTube, and so on. While all were instances of the so-called social web and thus all "harnessed user contributions," Wikipedia was distinguished by other important characteristics: rather than merely "harness" users, Wikipedia was constituted almost entirely of them; Wikipedia was not profit-seeking and its outputs were not turned into commodities; Wikipedia operated under a unique copyright license borrowed from software cultures,[4] which allowed its contents to be modified and reproduced without seeking permission or remunerating the prior "authors"; and, finally, Wikipedia was a massive attempt to make a single, unified thing. To be sure, there were many different components to the project—it was, to use Yochai Benkler's terminology, highly modular and granular (2006, 100)—but these components were all organized, in one way or another, around the vision of making an encyclopedia. Wikipedia was a very new attempt at a very old genre—one with a rich and varied history.

As historical artifacts, encyclopedias have regularly offered great insight into the periods in which they were written. They tell us about what constitutes knowledge at a particular time as well as how the various bodies of knowledge were thought to relate to one another. Encyclopedias also tell us how knowledge is to be received, how it is to be read, and what is at stake in the acquisition of its contents. And as any scholar of the Enlightenment would attest, encyclopedias are also highly politicized artifacts. The history of the famous French Encyclopédie, for example, was filled with conflict, turmoil, and political intrigue. At times, its production was officially suspended and its contents suppressed, and the possibility of these actions haunted the project for much of its life. The Encyclopédie's central and somewhat tragic figure, Denis Diderot, was himself imprisoned and regularly surveilled by authorities.

Encyclopedias are also rich sources of organizational, economic, and media-archaeological (Huhtamo and Parikka 2011) insight. Robert Darnton's classic work, *The Business of Enlightenment: A Publishing History of the Encyclopédie, 1775–1800* (1987), provides an excellent case in point. Darnton focuses on the complex realities of disseminating "Enlightenment" through the publishing history of the Encyclopédie. Books like the Encyclopédie, he reminds us, were simultaneously "products of artisanal labor, objects of economic exchange, vehicles of ideas, and elements in political and religious conflict" (1), and his text speaks to all of these things. More than this, though, Darnton shows the value—perhaps even the necessity—of thinking them all

---

4. At the time, they used the GNU Free Documentation License, but would switch in 2009 to the Creative Commons Attribution–ShareAlike License.

together, of showing their interrelations. His work, which concentrates on the period after compilation and editing of the Encyclopédie are largely finished, brings together detailed accounts of bookmaking; legal and geopolitical considerations around publishing; strategies of distribution, subscription, and payment; and the politics and economics of different editions and publication types (folio, quarto, octavo, and so on), among other things. Darnton uses these considerations to make a series of arguments about things like the readership of the text; the role of copyeditors and advertisers in relation to the form and stability of the text; the value of the materiality of the book during this period (as opposed to its contents); the realities of work in the printing houses; the role of piracy in the dissemination of Enlightenment; and the ruthlessness of publishing in the period of "booty capitalism." Most important, Darnton argues that the entirety of relations under the name "Encyclopédie," and not just its written contents, must be understood as an instantiation of Enlightenment:

> By studying how the Encyclopédie emerged from the projects of its publishers, one can watch the Enlightenment materialize, passing from a stage of abstract speculation by authors and entrepreneurs to one of concrete acquisition by a vast public of interested readers. (520)

The Encyclopédie was more than a book; it was a microcosm. On the one hand, it was a work of absolute specificity, distinguished from all other books of the period and all of its historical counterparts. On the other hand, it was equally an instantiation of ideas, values, techniques, and procedures that were widespread:

> A whole world had to be set into motion to bring the book into being. Ragpickers, chestnut gatherers, financiers, and philosophers all played a part in the making of a work whose corporeal existence corresponded to its intellectual message. As a physical object and as a vehicle of ideas, the Encyclopédie synthesized a thousand arts and sciences; it represented the Enlightenment, body and soul. (522)

In much the same way as the Encyclopédie served as fodder for scholars of the Enlightenment period,[5] I was interested in what Wikipedia could tell us about the conditions of knowledge production in the networked present. What constitutes our encyclopedic knowledge and what are its limits? How is the Wikipedia project organized, not only in terms of its taxonomy of

---

5. Other contenders roughly around this period include John Harris's *Lexicon Technicum*, Ephraim Chamber's *Cyclopaedia*, the *Encyclopaedia Britannica*, or the *Encyclopédie méthodique* (see Yeo 2001).

knowledge, but also in terms of the creation and editing of articles, the distribution of tasks and hierarchy of contributors, and the entire arrangement of the project's technical components? In other words, how is it organized as a project? How do people work together, make decisions about what stays and goes, about what is good and bad? What are the key principles that flow throughout the project, leaving their mark on everything else? What happens when previously committed contributors become disenfranchised? And finally, what are the interrelations between all these things? Is it possible to think them together?

Starting small, the first thing I focused on was how people work together in Wikipedia. How are new contributions made to the project? How is work organized? These are the questions I explore in detail in chapter 2. I quickly noticed that considerations of these questions were entirely dominated by the related notions of participation and collaboration. After all, Wikipedia was the encyclopedia "anyone can edit." Henry Jenkins would use Wikipedia as a key example of his "participatory cultures" (2006a); Axel Bruns would use it as an example of "open participation" (2008a); Clay Shirky would use it as a model of "collaborative production" (2008); Tapscott and Williams would name their book about "mass collaboration" *Wikinomics* (2006); and Joseph Reagle titled his book about Wikipedia *Good Faith Collaboration* (2010). But it wasn't simply that Wikipedia was being described in these terms; rather, it was being held up as the ideal instantiation of them. Wikipedia was the shining star of participation and collaboration. But even though these notions were dominating the discussion about contributing to Wikipedia—and about the future of working together in general—I found most of them rather thin on the actual details of working together to create articles.

The mere fact of participation seemed largely meaningless when it was clear that although anyone with an Internet connection could technically edit, there were absolutely no guarantees that an edit would stick. Collaboration, which was more often deployed to capture the quality of actually working together (under the general conditions of participation), was most often described as nonhierarchical and self-organizing or as a spontaneous coming together of peers. It was a mode of working together without bosses and without the chains of command that defined old institutions (or at least with "benevolent" leaders whose authority was utterly contingent on the will of those following). Despite such claims, it seemed obvious that there were all kinds of hierarchies to be found, the most obvious of which lay with the contributions themselves: between ones that were accepted and ones that were deleted. Equally obvious was the large body of rules for determining the status of a contribution. But something in the makeup of these terms, collaboration and participation, made it difficult to capture the political aspects of working together, that is, of how some contributions were chosen at the expense of others and how these decisions marked contributors in asymmetrical ways. To be sure, there was the odd controversy, but in the language of collaboration and participation Wikipedia's mode of working together was largely devoid of serious antagonism. To use Chantal Mouffe's (2005) term, participation and collaboration seemed to exist in a *postpolitical* space as an answer to or resolution of prior politics.

This way of thinking about Wikipedia, both as a model for the future and a model without a politics, and therefore as a future without politics (politics conceived in terms of the necessary existence of agonistic and antagonistic encounters) was similarly present when I began to consider other aspects of the project. When looking at governance and authority in Wikipedia, I came across notions like merit (or meritocracy), charisma (in the Weberian sense), and the elusive ad-hocracy. Likewise, when I investigated a very messy event in Wikipedia's history, which involved a bunch of disillusioned people leaving to start a competing project, I found it described in the language of forking—one that suggests that there are minimal losses involved in leaving a project and starting a competitor. I thus came to focus first and foremost on this problem of the political. What kind of politics were attributed to the general condition of participation, the working together of collaboration, the organization of ad-hocracy, and the exodus of forking? The answer: openness. As Obama's memorandum made clear, openness is what orientated these elements into a coherent politics. But openness didn't seem to "solve" any of the problems I had encountered with its subconcepts: the problem of a future without politics. In fact, openness seemed to further obfuscate the question of politics; it was explicitly political and postpolitical at the same time.

My first encounter with political openness, then, did not come about via an involvement in free and open source software or from a search for progressive politics. It came about through an interest in the organization of an encyclopedia project and the realization that the two were intimately connected. To think politically about Wikipedia, one had first to pass through openness. Conversely, for those interested in political openness and its allied notions, Wikipedia was its model par excellence.

The first chapter of this book therefore begins not with a discussion of a particular aspect of Wikipedia, but with the problem of political openness in general. The chapter is largely historical and cartographic. I make the argument that contemporary political openness cannot be understood without reference to debates about open systems in the 1980s and the emergence of free and then open source software shortly after. It is largely from these cultures that openness became aligned with notions like collaboration, participation, and exit-as-forking. Although a major aim of the chapter is to establish the prominence of openness and connect it to developments in software and computer systems, I do not suggest that openness originates from this period fully formed; nor is the point solely to announce its (re)emergence. Rather, while openness passes through these technocultures, undergoes significant transformations, and then continues to flow into other domains, it also precedes them. Openness has a longer history; it has been deployed to address different if perhaps not entirely unrelated political concerns and it has been allied with other notions. Some of these, I suggest, such as decentralization and the logic of the market, continue to bear upon its contemporary character, while others have receded. As noted, though, the focus is not on the emergence or re-emergence of openness per se, but on the problem of openness. Thus, the chapter not only provides a cartography, but also investigates the function and logic of openness, both past and present. As a historical reference point, I rely on Karl Popper's classic *The Open Society and Its Enemies* (1962), a work that reframes the entire history of political philosophy as well as the realities of World War II in terms of the open. I build a critical account of openness through a reconsideration of this text in relation to more recent appropriations.

If the first chapter is critical and cartographic, the remaining chapters narrow in focus. While each chapter extends the critical inquiry, each also contributes positively to a revised method of political description for projects organized around the logic of openness. As noted above, chapter 2 considers working together to produce encyclopedia entries and begins with an account of participation and collaboration. The chapter comes to focus specifically on how collaborative work is organized into desirable and nondesirable contributions. This question of "sorting collaboration" is explored through two detailed case studies. The first covers the attempt by Nathaniel Stern and Scott Kildall to create an experimental art piece on Wikipedia titled "Wikipedia Art." While these artists use Wikipedia Art to comment on the possible circularity of Wikipedia's citation mechanism, I use it to illuminate Wikipedia's article-deletion process, focusing in particular on the different modes of argumentation mobilized by participants. The second case study follows a lengthy and heated debate over the depiction of images of Muhammad. This second case study shares similarities with the first, but plays out on a different scale and in relation to a different set of "mediation" procedures. Through these case studies, I argue that collaborative work in Wikipedia cannot avoid what I call a politics of the frame. I draw on the work of Gregory Bateson (1972) and Erving Goffman (1974) to develop this notion of the frame, and then on Jean-François Lyotard's "differend" (1988) and Bowker and Star's "boundary objects" (1999) to further articulate its political dimensions.

In chapter 3 I turn to the major organizing principles of Wikipedia—the principles that define its frame. The chapter begins with a consideration of the literature on Wikipedia's mode of governance and comes to focus on the counterposed notions of ad-hocracy and bureaucracy. While I don't suggest that Wikipedia is either of these things, I use aspects of Max Weber's original account of bureaucracy to explore the organizing force of the project's framing statements. The chapter considers Wikipedia's hierarchy of policies and guidelines (with a detailed critical account of the neutral point of view policy); the organizing work of its "apparatus of material implements" (Weber 1958, 197), including a discussion of software "bots"; as well as a consideration of user hierarchies (e.g., anonymous users, administrators, stewards). While I do not develop anything like a full theory of agency, this last consideration does lead to some general observations about the capacity of contributors to act within the Wikipedia formation.

Chapters 4 and 5 are oriented around the notion of forking, which, as I noted earlier, is a mechanism for exiting an open project. Although forking might take place without major conflict or controversy, it plays a crucial role when such conflict is present. Chapter 4 places the discourse of forking within a broader tradition of political exit. Forms of leave-oriented political action, from exit and exodus to revolution, may serve one of two functions. On the one hand, they represent an actual technique of the subjugated: a way to transform the existing state of things. On the other hand, they function in an entirely different manner, as a mechanism of legitimization. In this sense, the ability to leave a form of political organization—when such ability is built into the very constitution of the organization—serves to legitimate the existing organizational form. If members of the organization are really unhappy, the logic goes, they are free to leave or overthrow it. In chapter 4 I begin by examining forking primarily as a mechanism of legitimization. I place forking in relation to other discourses of political exit and, drawing in particular from the writings of Albert Hirschman (1970), I illuminate the novelties of forking within this tradition. The second part of the chapter introduces an attempted fork of the Spanish Wikipedia that took place in early 2002. Through a consideration of the "Spanish Fork," that is, of an actual attempt to execute the political technique of forking, I reflect upon the relationship between forking-as-technique and forking-as-legitimating-mechanism. Rather than dismiss the political discourse of forking outright, the chapter tries to account for the ontological conditions that make forking both intelligible and possible.

Chapter 5 continues with the Spanish Fork of Wikipedia. Instead of focusing on forking as a mechanism of legitimization or as a technique of political exit, this chapter reframes forking as an instance of controversy. I look closely at a debate that played out on one of Wikipedia's early discussion lists over a two-week period in February 2002 to explore how controversies emerge and develop, how different statements gather or lose force and how they face off against one another, as well as the kinds of representational techniques available to participants in the controversy. Throughout the analysis, the chapter engages closely with the political ontology (of statements) developed in Bruno Latour's early writings on controversies in science (1987). A secondary aim of the chapter is therefore to translate Latour's insights from scientific practice into the domain of network cultures—and "open projects" in particular—and to stress their political character. It seeks to affirm the political character of Latour's writings (and actor-network theory more generally) beyond the typical emphasis on the mere fact of materiality, the nonhuman, or the "parliament of things," to focus instead on the agonistic ontology of statements.

Although the following chapters bring in many characters—pillars, principles and guidelines, artists, deletion procedures, mediators, sysops, anonymous users, project founders, disgruntled forkers, bots and bot approval groups, rival encyclopedias, and historical precedents, to name a few—I do not claim that Wikipedia represents the "body and soul" of openness, as Darnton writes of the Encyclopédie and the Enlightenment. As noted, others have already made this claim often. Nor do I claim comprehensive coverage of Wikipedia. Instead, I dive into selected events and cases from Wikipedia's history chosen specifically for what they reveal about the organizational qualities of the project in relation to openness. What I am searching for is a very specific kind of politics, an organizational politics, and one that leaves other possible political trajectories aside.

There is no discussion, for example, of the Wikipedia "blackout" of early 2012, where the site went offline in protest of the proposed SOPA (Stop Online Piracy Act) and PIPA (Protect IP Act) legislation. This was perhaps the first time that the project saw itself explicitly as a political actor in the world and used its value as a public good to promote a political agenda. The "blackout" was certainly a political moment, but it wasn't directly about the internal organization of the project.[6] I also don't focus in detail on the activities of the Wikimedia Foundation, which oversees various aspects of the project

---

6. To be sure, it did pose the interesting question of how a distributed and supposedly leaderless organization can mobilize itself to become a coherent political actor.

as it interfaces with the rest of the world (see Fuster Morell 2011; Chen 2011). There are legal, economic, administrative, and meta-governmental functions of the Foundation that undoubtedly raise interesting questions about the ideals of openness. In a similar vein, I don't consider Wikipedia's function within the wider web ecology (such as its generally symbiotic relationship with Google Search) or the many instances where outside forces have tried to shape the encyclopedia for their own gain. Instead, I want to respect as much as possible what might be called the "magic circle" of openness.

Finally, I do not take well-established categories of political thought as a starting point (e.g., age, race, class, and gender) and then apply them to Wikipedia. In this regard, a much-publicized concern with Wikipedia is its longstanding and rather severe gender imbalance.[7] For some, this fact alone is enough to dismiss Wikipedia's openness as the emperor's new clothes. It is also clear that the project has made negligible inroads toward rectifying this imbalance since its "discovery." I think there are several main reasons for this:

(1) While open projects can be found to contain "gender gaps," the language of openness itself is compatible with the conditions of gender equality. Remember, "anyone can edit." The project encourages participation and collaboration, and if people are really unhappy, for any reason, they can always create a fork. So there are a series of possible responses that can fend off a gender critique based on empirical grounds: "it's not our fault that there are more male contributors because anyone can edit"; "if the situation were that bad, people would fork the project and realize their own gender-balanced vision." A more extreme position could even place the blame with the nonexistent female contributors for "not participating." (2) Even though a gender gap exists, it is not possible to deal with it directly (through policy intervention, for example) without contradicting openness. If Wikipedia were to declare that all or even 50 percent of new editors had to be female, this declaration would explicitly go against the notion that "anyone can edit." (3) Whenever specific sexist activity is identified, in fact whenever anything at all is identified that is clearly wrong or damaging to the project and made public, Wikipedia's openness is such that it is usually quickly corrected.

In April 2013, for example, novelist Amanda Filipacchi discovered that female novelists were being removed from the "American Novelists" category

---

7. According to a 2010 survey, only 12.64 percent of 53,888 people who identified as project contributors were female; R. Glott, P. Schmidt, and R. Ghosh, "Wikipedia Survey—Overview of Results," 2010, http://www.wikipediasurvey.org/docs/Wikipedia_Overview_15March2010-FINAL.pdf.

and placed instead into an "American Women Novelists" subcategory, while no such equivalent category had been created for men (who remained in the main "American Novelists" category). Filipacchi drew attention to this fact in a *New York Times* article titled, "Wikipedia's Sexism toward Female Novelists" (2013). In the days that followed, there was a spike in editing activity around these categories and their related pages. This included some predictably juvenile behavior (Filipacchi's own page even came under attack), as well as the revelation of some very bizarre "revenge editing" undertaken by an editor, "Qworty," who was at the center of the controversy (see Leonard 2013a, 2013b, 2013c). After the dust had settled, however, the female authors were restored to the main "American Novelists" category, and a subcategory for "American Male Novelists" had also been created. While these events can be very off-putting and reveal both naïve and malevolent forms of discrimination, and while they should rightfully be addressed directly (as Filipacchi and others did), this is not the approach I will take. I am more interested in how Wikipedians and other project defenders can reconcile such glaring issues with the ideals of openness and carry on unperturbed. What line of reasoning makes this possible? In 2011, for example, former director of the Wikipedia Foundation Sue Gardner was quoted in another *New York Times* piece as wanting to increase the female contributor base to 25 percent by 2015 (Cohen 2011). According to the author of that piece, Gardner's effort "is not diversity for diversity's sake." Gender equality was not the rationale. What was? Gardner: "Everyone brings their crumb of information to the table.... If they are not at the table, we don't benefit from their crumb" (cited in Cohen 2011). The gender gap has been "positively" transformed. Women are now information-storage devices whose inclusion is desired in order to increase the resource base of the project. Women add to the pool—the market?—of available information. While Gardner clearly wants to address the "gender gap," she does so through a different kind of logic, which is clearly privileged over gender per se. It is this logic that concerns me.

Rather than bring sociological categories to bear on the project in order to show its failings, I set myself the more difficult task of showing how the project is necessarily involved in the sorting of knowledge, categories, and statements, which all have identity effects. It is not merely that the occasional gender issue pops up and needs to be addressed. All forms of organization are generative of differences and such differences can easily provide the basis for new (and old) types of subjugation. Thus, instead of focusing on gender from the outset, the challenge of this more immanent method is to understand the specific mechanisms, the procedures, the forms of reasoning, and so on, from which new categories of exclusion and subjugation emerge and old ones either fall away or are buttressed. It is the production, rather than the outcome, that is the focus of my attention.

The challenge of this book is to capture something of the organizational politics of Wikipedia and to rub these up against the language of openness, revealing its tensions, contradictions, subjugations, invisibilities, and lines of force. Most important, the challenge is to build a set of concepts and techniques of political description, borrowed, revised, and brought together from a range of historical and contemporary resources, from which it becomes possible to speak coherently back to openness.

1

Open Politics

> Most think about these issues of free software, or open source software, as if they were simply questions about the efficiency of coding. Most think about them as if the only issue that this code might raise is whether it is faster, or more robust, or more reliable than closed code. Most think that this is simply a question of efficiency. Most think this, and most are wrong. . . . I think the issues of open source and free software are fundamental in a free society. I think they are at the core of what we mean by an open society.
>
> LAWRENCE LESSIG (2005, 260)

> One approach to understanding the democracy of the multitude, then, is as an open-source society, that is, a society whose source code is revealed so that we all can work collaboratively to solve its bugs and create new, better social programs.
>
> MICHAEL HARDT AND ANTONIO NEGRI (2004, 340)

"The open" has become a master category of political thought. Such is the attraction, but also the ambiguity of openness, that it appears seemingly without tension, without need of clarification or qualification, in writers as diverse as the liberal legal scholar Lawrence Lessig and the post-Marxian duo Michael Hardt and Antonio Negri. Every political position worth its salt, it seems, must today pledge allegiance to this strange and relatively new political concept. The epigraphs above are indicative of a development that forms the basis of this chapter: the re-emergence and repoliticization of openness in relation to a set of developments specific to the realm of software. In the first epigraph, Lessig is looking back, trying to connect open source and free software to an already existing notion of open politics, the open society. Hardt and Negri—who, it must be said, are a long way from home on this matter—are looking forward, trying to establish a connection between the really existing practices and logic of open source software and their yet-to-be realized "democracy of the multitude." As does Lessig, I begin this chapter by connecting back, by revisiting a founding figure of open thought, Karl Popper. I trace what might be called the second coming of the open, through debates about open systems and open software in the 1980s and '90s, to the generalization and proliferation of openness in network cultures evidenced by such notions as open access, open education, and open communities, and, finally, to the reemergence of the open in institutional politics and related writings.

My purpose is not to pin down the meaning of openness, or to moralize this notion, but to trace its proliferation and to consider how it functions in contemporary cultures, in the writings of Popper, and in relation to competing and supporting concepts. Through a reconsideration of the open in the writings of Popper, I finish by outlining some concerns for contemporary proponents of open politics—a task that I consider crucial as the open is increasingly used to "look forward."

The Open Society

Karl Popper was not the first to write about the concept of openness, or even of the open society. Henri Bergson had already used the term "open society" in his work of moral and political philosophy, *The Two Sources of Morality and Religion* (Bergson 1935). Likewise, Austrian biologist and pioneer of general system theory Ludwig von Bertalanffy had described the living organism as an open system, distinct to the closed systems that characterize machines (von Bertalanffy 1950, 1960). Von Bertalanffy described open (and therefore living) systems as those that maintain themselves through the constant exchange of materials with their environment, a process that also entailed a "change of components" of the system itself (von Bertalanffy 1950, 23). In both these authors we already see developments that have today become commonplace: the conjunction of open and system, or more specifically the emergence of openness as a quality of a system, and the moralization and politicization of open and closed. However, it wasn't until Popper wrote *The Open Society and Its Enemies* (1962) while in exile in New Zealand during World War II that the political notion of the open gained a wider resonance.[1]

In two volumes Popper rewrites the history of political philosophy, and also lived political conflict, around the concept of openness. He locates the origins of his version of "the Open Society" in the "breakdown of Greek tribalism" (1962, 183), culminating in the Peloponnesian War (circa 431–404 BC) between the Delian League, headed by Athens, and the Peloponnesian League, led by Sparta. Interwoven with this history is a detailed critique of

---

1. While Popper's notion of openness is the focus below, it is worth pointing out from the start that the reference to von Bertalanffy is not unrelated to Popper's own intellectual trajectory. A certain mode of "open" thought emerged in Vienna in the 1920s and '30s, which von Bertalanffy popularized in biology but which equally captured the attention of Friedrich Hayek in economics and Popper in political philosophy. Indeed, Hayek was friend to both von Bertalanffy and Popper and the mutual influence of them in his work is easily detected.

Plato, whose political philosophy Popper argues is strongly marked by these events. For Popper, Plato is the first major proponent of "closed societies" (although he notes Plato's indebtedness to Hesiod and Heraclitus). Popper describes the war as a pitting of the old tribal form of society, Sparta, against the newly emergent open society of Athens, characterized by its democratic and equalitarian political organization (183).

Plato is depicted as a brilliant but misguided thinker whose experiences of the war (especially the execution of his mentor Socrates) lead him to build a totalitarian and reactionary political philosophy. This philosophy, Popper writes, is built on the principle that virtually all change is bad and society, which is always "in flux," is therefore in a state of deterioration. In opposition to this state of flux, Plato posits an original ideal form of society existing in ancient history, highly stable and resistant to change, from which the current imperfect society is derived. This original state equates to the *theory of forms or ideas* that underpins Plato's philosophical thought: the original tribal society is the ideal, whereas the actually existing society, with all its problems, is the inferior and degraded version of this form. (I leave aside the interesting dilemma of how the perfect form is able to degrade.) In the battle between Athens and Sparta, therefore, the older, "tribal" Spartan social structure is considered more desirable as it is closer to the ideal form, while the Athenian democracy represents radical change and therefore degeneration. It is around this notion of negative change and the ideal ancient Greek tribal form that Plato writes *The Republic* (1974), and which, Popper reminds us, is more accurately translated as The State. *The Republic* describes a society where all change is arrested.[2] The social is organized around three classes—rulers, auxiliaries, and producers—each with highly specific roles. The whole social edifice—education, law, reproductive norms, and so on—is designed to maintain this strict demarcation and rigid order. There is no "cross-breeding" between the classes and social interaction is avoided.

Philosophy, conceived as the perception of ideal forms, emerges in Plato's thought as the bridging device from the status quo to this ideal state. As the famous "simile of the cave" passage reminds us, Plato posits the philosopher as the only actor able to see true knowledge—the light of the sun as opposed to the shadow puppets on the cave wall—and thus as the only individual qualified to determine how a society should be organized. Such enlightenment also distances the philosopher from the desires and indulgences of

---

2. Popper gives this account of Plato's position: "The state which is free from the evil of change and corruption is the best, the perfect state. It is the state of the Golden Age which knew no change. It is the *arrested state*" (1962, 1:29).

everyday life and thus makes them even more suitable rulers of society—so-called philosopher kings.[3]

Popper critiques Plato on multiple grounds, but the overall argument can be summarized as follows: Plato claims to possess a kind of true knowledge, the knowledge of forms. This knowledge provides the general laws of history (what Popper elsewhere calls *historicism*) and at the same time positions the philosopher as the only person able to steer society in the right direction (because of the knowledge the philosopher possesses about how things should be). All decision-making capacity is removed from everyone except the philosopher, who decides in the most disinterested fashion what is right for all; armed with the knowledge of history, with its ineluctable laws, the philosopher is compelled to become a *social engineer*. Deprived of any capacity to choose due to the reification of all roles and duties, coupled with the subjugation of nonphilosophical knowledge—the mere "knowledge of shadows"—the individual is effectively denied agency. Individuals in Plato's perfect society cannot change or challenge roles; their path is fixed and their (nonphilosophic) knowledge emptied of value.

Although Plato's influence remains crucial, Popper's critique of closed thought and politics is then extended well beyond the writings of Plato. Any political philosophy based on unchallengeable truths—such as the discovery of the *laws of history*—that provide definite and rigid future programs and where individual will is always subordinated to these larger truths, is described in the language of the *closed society*. For Popper, the three most important philosophers in this tradition are Aristotle, Hegel, and Marx. Aristotle is largely dismissed as Plato's mouthpiece, with the exception that he puts a positive spin on Plato's theory of forms: rather than constantly degrading, the state is positioned as heading toward an ultimate end, toward perfection. Aristotle is nevertheless important for Popper because his biologically influenced teleological thought is taken up by Hegel, which in turn informs German Nationalism through the notion of the destiny of one race (the most perfect) to rule all others, as well as Marx's laws of class struggle and the destiny of the proletariat. Thus, Plato is significant not only as the first closed thinker or "enemy of the open society," and not just because he influenced these key historical figures, but because it is his political philosophy that informed two of the three major competing political programs during World War II, fascism and communism. In Popper's time, therefore, fascism and communism are the modern manifestation of the closed society,

---

3. Plato writes, "Wouldn't his eyes be blinded by the darkness, because he had come in suddenly out of the sunlight?" (1974, 243).

while capitalism and the democratic institutions affiliated with it represent the open society.

The summation of Popper's thought is a rearticulation of existing political concepts (democracy, fascism, communism), of key historical figures of political philosophy and their writings (Plato, Aristotle, Hegel, Marx, and others), and of lived conflict (the Peloponnesian War, World War II) around the new master categories of the open and the closed. In this new politics of the open/closed the fate of a nation and its people, or the class inequalities produced by capitalism, are no longer the primarily concern. The question is no longer about identity, race, or class, but about whether or not a social program, that is, a set of knowledges and related practices, is able to change. Social programs based on unchallengeable truths, so-called laws of history or of destiny, emerge as the fundamental enemy and what might be considered radically different political programs in a different frame of analysis, communism and fascism, are made equivalent. The positive side of this political equation, the open society, is one where totalizing knowledge is necessarily impossible.[4] Openness is necessary because no one can know for certain what the best course for society might be from the outset, and at the same time it is assumed that openness provides the best possible conditions for producing knowledge and therefore making better decisions.

I return to Popper and the open society below, but first I want to map the re-emergence and rearticulation of openness, beginning in software cultures, through to network cultures and more traditional political institutions. I want to demonstrate the significance of openness by highlighting its proliferation and also to the way it is increasingly held as the highest political ideal.

From Systems to Source, or, How We Became Open (Again)

By the 1980s, the United States was under the sway of neoliberalism. The organizational philosophy of "competition" had seemingly defeated the socialist desire for "centralized planning" in the socioeconomic ideology wars. The

---

4. It is at this intersection of social organization and truth that the connection between Popper's philosophy of science and his political philosophy is most apparent: For Popper, a scientific fact is by definition challengeable and science proceeds by disproving pre-existing facts. Science is precisely not about the establishment of everlasting truths. This knowledge-based critique, that is, of the impossibility of absolute certainty and therefore perfect knowledge, also parallels those made by Friedrich Hayek against central planning economies. This connection between the open society and neoliberal economic thought is of central importance and is considered below.

main philosophical argument used to justify what would become the neoliberal agenda was provided by Friedrich Hayek (1944) and resonates strongly with Popper's notion of the open society.[5] Hayek argued that the knowledge of how a society should be organized and which direction it should take is beyond any one individual or group and can never be known with certainty. Because of this, any attempt at centralized planning (i.e., socialism, communism, fascism), which is founded on exactly the assumption that what is best for all society is directly knowable, is likely to make bad decisions that only satisfy a small group. For Hayek, giving one group the ability to make decisions for the whole results in the overall reduction of liberty and the advent of totalitarianism. Instead, Hayek suggests, once society reaches a certain complexity, only a decentered mode of organization, where competing ideas and practices can interact and adjust in relation to change, can ensure liberty:

> It is only as the factors which have to be taken into account become so numerous that it is impossible to gain a synoptic view of them, that decentralisation becomes imperative. . . . Decentralisation has become necessary because nobody can consciously balance all the considerations bearing on the decisions of so many individuals, the co-ordination can clearly not be affected by "conscious control," but only by arrangements which convey to each agent the information he must possess in order effectively to adjust his decisions to those of others. (Hayek 1944, 51)

The precise form that decentralization takes is competitive markets. Such markets theoretically enable many individuals to shape society through the sale and purchase of commodities and thus with no "conscious control." Freedom is therefore intimately tied to economic freedom, to the freedom to sell commodities, including human labor, in a market. But the argument for economic freedom derives from a more fundamental critique of knowledge

---

5. As noted above, it is well known that Popper and Hayek were friends and intellectual allies. In the 1978 interviews at UCLA, for example, Hayek comments on the affinity between his own ideas and Popper's: "Now, the relation is, on the whole, curious. You see, Popper, in writing already The Open Society (and Its Enemies), knew intimately my counterrevolution of science articles. It was in these that he discovered the similarity of his views with mine. I discovered it when the Open Society came out. Although I had been greatly impressed—perhaps I go as far back as that—by his Logic of Scientific Discovery, his original book, it formalized conclusions at which I had already arrived." In other sections of the interviews he repeatedly speaks favorably of Popper's writings and their friendship: "Ever since I have been moving with Popper. We became ultimately very close friends . . . on the whole I agree with him more than anyone else on philosophical matters" (F. A. v. Hayek, "The Hayek Interviews: Alive and Influential," http://hayek.ufm.edu/index.php/Main_Page).

and centralization. Thus, the critique of totalitarian knowledge put forward by Popper and shared by Hayek is translated into government and economic policy to justify competitive, market-based forms of organizing society.[6]

With these larger changes in the theory and practice of governance taking place in the background, important new contests over openness arose in computer cultures, specifically around the notions of open systems and, soon after, software. These contests were seemingly far removed from Hayek-inspired neoliberal agendas but, as we shall see, arguments made by Popper and Hayek at the level of philosophy and economics are isomorphic with the ones that played out in computer cultures. In regard to systems, software anthropologist Christopher Kelty has covered the early debates about openness that played out around the UNIX operating system as well as the TCP/IP protocols. He describes these debates as at once technical and moral, "including the demand for structures of fair and open competition, antimonopoly and open markets, and open standards processes" (2008, 144). In the open systems debates, the battle for openness is not against closed forms of knowledge, à la Popper, but against proprietary standards—what might be described as closed infrastructures. I will not recount the history of these debates in detail, which has already been done very well by Kelty. Instead, I will focus on one story that developed throughout this period: the birth of free software and the challenge of open source. I focus on this story because it surpasses notions of openness in open systems and captures both the lived experience and the contested distributions of agency in software cultures. It reveals how competing mutations of liberalism were aligned with new legalities and modes of production and, most important, how all these developments would redefine and re-energize political openness.[7]

In 1980 a group of programmer-hackers at MIT, including a young Richard Stallman, were confronted with a problem: the AI Lab they were working in had received a new Xerox 9700 laser printer.[8] As the printer station was

---

6. For an intellectual and organizational history of neoliberalism, see *The Road to Mont Pèlerin: The Making of the Neoliberal Thought Collective* (Mirowski and Plehwe 2009). See also David Harvey's *A Brief History of Neoliberalism* (2005).

7. The most detailed version of this narrative that I have found is Sam Williams's *Free as in Freedom* (2002), although additional fragments can be found all over the web.

8. It is worth noting that around this time the distinction between "user" and "programmer" were not as firmly set as they would become and the term "hacker" was only beginning to be taken up by fear mongers (governments, companies, and film studios). Instead, the term "hacker" was akin to "tinkerer"—people who were interested to see how things work in order to understand, improve, or alter them. Interestingly, in his early commentary on hackers Steven Levy writes about a "common element," a "common philosophy" to all hackers: "It was a phi-

located on a different floor to the majority of people who use it, the young Stallman had written a small program for the previous Xerographic printer that electronically notified a user when his or her print job was finished and also alerted all logged-in users when the printer was jammed. This required some minor modifications of the Xerographic printer's code. When the new Xerox machine arrived, Stallman intended to make similar program modifications. But curiously, this new machine, which was offered to the lab as a "gift" from Xerox, did not arrive accompanied with a document containing the printer's (human readable) source code.[9] Without the source code, no modifications could be made to the Xerox and the tyranny of distance between the people on one floor and the Xerox on another would be felt once more! Stallman grew increasingly suspicious of this act, or rather nonact, by Xerox. Up until this point it was common courtesy to supply the source code along with any program that entered the laboratory. When it became clear that the source code was not going to appear on its own accord, Stallman decided to track down the original programmer to ask for the source code personally. On confronting the programmer, he was told that he could not have a copy of the source code and moreover, that the programmer had signed a nondisclosure agreement (NDA), which at the time was a complete novelty in the field of software. Recalling the incident, Stallman notes, "It was my first encounter with a nondisclosure agreement, and it immediately taught me that nondisclosure agreements have victims. . . . In this case I was the victim. [My lab and I] were victims" (cited in Williams 2002, 11). It was after this encounter with the Xerox, so the story goes, that Stallman famously declared, "All software should be free." Not (only) in the sense of "free to use" or "free to distribute," but in that greater sense of "free to change, modify, rewrite,

---

losophy of sharing, *openness*, and getting your hands on machines at any cost—to improve the machines, and to improve the world" (1984, 7; emphasis added).

9. While all operative computer code is eventually translated into long streams of binary notation (ones and zeroes), the type of code read and modified by humans is referred to as source code. For the time being, such code can be thought to exist at the threshold of computational and semantic/linguistic communication. While binary notation is technically decipherable (and indeed, the first programmers operated with this notation alone), it is immensely difficult and entirely unrealistic in contemporary software environments. Source code, written in an array of languages (e.g., Java, C++, Python, Basic), allows a programmer to understand and modify a program, with the changes made on this "abstract," source layer eventually translated through various code processes (compilers, assemblers) into binary. Withholding the source code is a key strategy of control that removes the capacity to interact with a program, thus limiting the kinds of agency possible in a given software environment. As I note below, this is also central to the commodification of software.

adapt"—in short, a freedom to reorganize and modify the algorithms that instruct the machines that populate our worlds.

After growing increasingly disillusioned with the effects of commodification as a mechanism of control, taking place both in his immediate environment and the wider software community, Stallman left the lab at MIT. His plan was to create an entire operating system (OS) that would not be subject to what he perceived as the *closures* of proprietary software. In 1983 he announced plans to create the GNU OS as part of a new Free Software Movement (FSM). The GNU OS was to be written from scratch using nonproprietary code.[10] In 1985 he set up a nonprofit corporation called the Free Software Foundation (FSF) to formally oversee and represent the movement. The most significant development during this period, however, was the creation of several unique copyright licenses designed to keep the outputs of the FSM "open." Although initially specific licenses were written for each new piece of software, in 1989 Stallman developed the GNU General Public License (GNU GPL) as a broadly applicable software license. These licenses are generally described as using the mechanism of copyright against itself, in that rather than restricting distribution through the creation of scarcity they use copyright to ensure an application or text can be accessed, made visible, dissected, and modified or "remixed" (Lessig 2008) by any user. The GNU licenses were not the only permission-based (as opposed to restriction-based) licenses, but the GNU GPL in particular was certainly the most progressive of its type; not only was any piece of software created under it accessible and modifiable, but the license states that any derivative of an earlier text/program must also adopt the same license. This was the legal mechanism that supported Stallman's desire to keep the outputs of his FSM "free" and the movement as a whole growing (because of the so-called viral nature of the license). The FSF would not only oversee the movement: Stallman also requested that any product making use of the license be signed over to the foundation, which would police any violations and take appropriate legal ac-

---

10. GNU ("gah-new") is an acronym of GNU's Not Unix. The Unix system that Stallman was eager to replace was initially created at Bell Labs by AT&T staff in 1969. While this OS was not commercial in the same sense of Microsoft Windows or Apple's OS X, it was still technically owned by AT&T. This ownership status, however, was complicated by the fact that many non-AT&T programmers had contributed to and helped improve the system. This murky legal situation is detailed in Chris Kelty's *Two Bits* (2008). Part of the desire to create a new OS from scratch, without borrowing even one line from the Unix system, was to avoid the possibility that a company might claim ownership over part of GNU and therefore exert control over it.

tion.¹¹ This story of Stallman and the Xerox circulates as the mythic origins of free software and establishes Stallman as its guru and prophet.

While Stallman proclaims that code is necessarily political, other programmers have attempted to uncouple this pairing (see Berry 2008, 147–87; Williams 2002, 136). Indeed, Gabriella Coleman (2004) has argued that the refusal to acknowledge their actions as political is one of the key characteristics of many software cultures. In 1998, for example, a group of high-profile programmers started the Open Source Initiative (OSI). The most vocal member of this group, Eric Raymond, viewed Stallman as a controlling idealist who focused too much on politics at the expense of technical excellence and efficiency. This strategic reframing of the question of code was designed to make free software business-friendly. The term "Open Source" was chosen to avoid the connotations surrounding Stallman's rhetoric of "free," which seemed less than appealing to profit-seeking enterprises, especially when attached to a product.¹² "Open Source" is well chosen as it foregrounds the technical dimension of these software practices—"this movement is about source code"—and conveniently sidesteps Stallman's political concerns. In order to achieve this distance from the FSM, the OSI had to generate its own licenses that effectively reversed the "viral" nature of the GPL. The challenge for these licenses—such as the Mozilla Public License—was to "balance" the requirement for companies to commodify software outputs with the increased potential for productivity made possible by involving outsiders and harnessing their "contributions."

The OSI also had its own gurus in Eric Raymond and Linus Torvalds. In 1997 Raymond published his first iteration of "The Cathedral and the Bazaar" (2000), a hugely influential musing on the production method of the developers of the Linux operating system—a project headed by Torvalds. The terms "cathedral" and "bazaar" are used to represent competing production methods. Of the cathedral method, Raymond writes:

> I had been preaching the Unix gospel of small tools, rapid prototyping and evolutionary programming for years. But I also believed there was a certain critical complexity above which a more centralized, a priori approach was

---

11. While many software projects have adopted the GNU GPL, not all have transferred "ownership" to the FSF.

12. It is important to point out here that companies such as IBM, Sun Microsystems, Intel, Novell, and Google are the major contributors to open source development. The romantic image of small groups of programmers banding together to outcompete big business is not historically accurate.

required. I believed that the most important software (operating systems and really large tools like the Emacs programming editor) needed to be built like cathedrals, carefully crafted by individual wizards or small bands of mages working in splendid isolation.[13]

The bazaar mode of production, found in Linux, emerges as the improbable yet superior other:

> Linus Torvalds's style of development—release early and often, delegate everything you can, be open to the point of promiscuity—came as a surprise. No quiet, reverent cathedral-building here—rather, the Linux community seemed to resemble a great babbling bazaar of differing agendas and approaches (aptly symbolized by the Linux archive sites, who'd take submissions from anyone) out of which a coherent and stable system could seemingly emerge only by a succession of miracles.
>
> The fact that this bazaar style seemed to work, and work well, came as a distinct shock. As I learned my way around, I worked hard not just at individual projects, but also at trying to understand why the Linux world not only didn't fly apart in confusion but seemed to go from strength to strength at a speed barely imaginable to cathedral-builders.[14]

Out of the FSM and the OSI emerge two competing mutations of liberalism. With Stallman lies the recognition that the creation of markets via the commodification of software actually reduces the capacities (or liberties) of individuals who use and modify it. The argument for open markets that played out in the open systems debates is extended to software itself. It is a liberal argument that fundamentally challenges the pre-existing liberal coupling of freedom and property. Openness is primarily understood as a techno-legal quality, whose opposite, as Kelty reminds us, "is not closed, but 'proprietary'" (2008, 143). With Raymond, on the other hand, the emphasis is not on commodification, but on the organization of production. Raymond's writings on the "bazaar" strongly parallel Hayek's arguments about the ideal organization of society described above, but applied at the level of individual contributions to specific software projects. The cathedral parallels the "centralized planning" critiqued by Hayek, while the bazaar emerges as a new liberal utopia: radically open to competing "agendas and ideas"; progress "at a speed barely imaginable"; and the miraculous emergence of a "coherent and stable system." The history of the OSI and the writings of Raymond in particular demonstrate how contemporary political openness came to be articulated

---

13. E. S. Raymond, "The Cathedral and the Bazaar," 2000, p. 3, http://catb.org/~esr/writings/homesteading/.

14. Ibid.

with a specific method of software development. While Stallman remained steadfast in his preference for the term "free" to describe his movement and its outputs, it was the business-backed "open source" and eventually just "open" that captured the minds of people beyond software cultures.

The Open Takes Flight

While software such as GNU/Linux-based operating systems, the Apache server client, and the Mozilla web browser have made free and open source software (FOSS) highly visible within software communities for many years, it is the *translation* of these ideas into new domains that is most significant.[15] Roughly since the turn of the millennium, various efforts have been made to create new products and processes based on the principles of FOSS. The material covered in this section is by no means exhaustive. I focus instead on a range of examples that demonstrate certain specific qualities of how openness has been translated. I begin with projects that name themselves "open" and thus explicitly interpret their activities in relation to openness. I then look at different projects that describe key facets of their activities in terms of openness, including activist groups that organize around openness, two "mainstream" entities (Wikipedia and Google), and, finally, different political writings and government initiatives that make use of the open. To be sure, there are many significant differences between all the examples covered, but my focus is on the fact that the same rhetoric is deployed by what are otherwise very different groups or organizations. It is the fact of diffusion that is most significant.

The most obvious translation of openness emerges from online projects or movements that explicitly name themselves as open. Within this category are broad movements or trends, such as open access, which is generally used to describe the making available of published content and especially scholarly, educational, and scientific materials. One example of open-as-open-access is the Open Humanities Press.[16] This initiative publishes academic monographs, but also acts as a kind of branding or certification mechanism for a series of online journals. Open Humanities Press has four stated principles that cover access, scholarship, diversity, and transparency, and a series of related goals, the last of which is to "explore new forms of scholarly collaboration."[17]

---

15. In 2001 Rishab Aiyer Ghosh coined the term FLOSS (Free/Libre Open Source Software) as an inclusive label for all these software projects.
16. Open Humanities Press, http://openhumanitiespress.org/index.html.
17. "About: Principles and Goals," http://openhumanitiespress.org/goals.html.

Other examples of open access include the Bentham Open project, which publishes "over 250 peer-reviewed open access journals"; unaffiliated journals such as Open Medicine, which makes its content "freely available for others to read, download, copy, distribute, make derivative works ('remix') and use with attribution" and whose stated mission includes the promotion of international "collaboration on health issues"; and finally, open access study materials, such as MOOCs (Massive Open Online Courses) and courses provided by participants in the Open Courseware Consortium, of which MIT's Open Courseware project is perhaps the best known.[18]

Closely related to these open access initiatives are projects that include an open access component, but also emphasize a broader or perhaps more "procedural" sense of openness. A good example here is the Open Knowledge Foundation. The Foundation "seeks a world in which open knowledge is ubiquitous and routine" and sees openness as having "far-reaching societal benefits." The Foundation states, for example, that politically "openness improves governance through increased transparency and engagement"; culturally, "openness means greater access, sharing and participation"; economically, "openness permits easier and more rapid reuse of material"; and "for science to effectively function, and for society to reap the full benefits from scientific endeavors, it is crucial that public scientific information be open." Because of these perceived benefits, the foundation supports and facilitates a range of projects, including the Open Data Commons, Open Shakespeare, Open Economics, Open Text Book, Open Milton, Open Knowledge Forums, Open Geodata, and Open Environmental Data. I have listed only the projects with "open" in the title, but there are numerous others.[19]

The strongest expression of *translated openness*, however, is to be found in the Open Everything movement (2010). Open Everything has a wiki that details its events and its function. The welcome page of the wiki states:

> Open Everything is a global conversation about the art, science and spirit of "open." It gathers people using openness to create and improve software, education, media, philanthropy, architecture, neighborhoods, workplaces and the society we live in: everything. It's about thinking, doing and being open.[20]

18. Welcome to Bentham Open Access," 2010, http://www.bentham.org/open/index.htm; Open Medicine, "Mission," 2010, http://www.openmedicine.ca/pages/view/Mission; Open Medicine, "Open Access," 2010, http://www.openmedicine.ca/pages/view/Open%20Access; MIT Open Courseware, 2010, http://ocw.mit.edu/.

19. Open Knowledge Foundation, 2010: "About Us," http://www.okfn.org/about; "Our Vision," http://www.okfn.org/vision/; "Projects," http://www.okfn.org/projects.

20. "Open Everything," 2010, http://openeverything.wik.is/.

FIGURE 1. Screenshot of Everything Open and Free mind map. *Source*: http://www.mindmeister.com/28717702/everything-open-and-free (accessed August 15, 2010).

Further down the page are a series of statements about the open:

> Open is changing the game. And, while Wikipedia and open source software offer great examples . . . we know that openness, collaboration and participation are spreading well beyond the realm of technology. . . . Where open is headed is huge. Open Everything gathers people who are charting this trajectory.[21]

Openness is conceived as a new mode of being, applicable to many areas of life and gathering significant momentum—"changing the game" as it were. Once again, this "spirit of open" is closely articulated with collaboration and participation. The Free and Open Everything initiative is also associated with the Foundation for P2P Alternatives, which functions "as a clearing house for open/free, participatory/p2p and commons oriented initiatives." The P2P Foundation has its own "Open Everything" directory, including a detailed mind map titled "Everything Open and Free" (see fig. 1), which attempts to comprehensively map and classify the dimensions of openness.[22]

Linked from the central "Everything Open and Free" hub is an array of different nodes, each of which covers a different dimension of openness, including Aspects of Openness (see fig. 2), Enablers of Openness, Infrastructures of Openness, Practices of Openness, Domains of Openness, Products of Openness, Open Movements, and Open Consciousness. The Open Everything project, together with the Everything Open and Free mind map,

21. Ibid.
22. Foundation for P2P Alternatives, 2010, http://p2pfoundation.net/The_Foundation_for_P2P_Alternatives. The full Mind Map is available at http://www.mindmeister.com/28717702/everything-open-and-free.

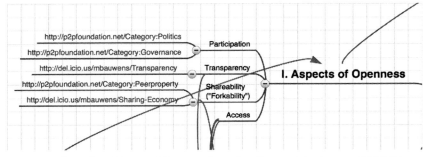

FIGURE 2. Screenshot of Aspects of Openness. *Source*: http://www.mindmeister.com/28717702/every
thing-open-and-free (accessed August 15, 2010).

represents an emerging desire to radically transform society around the concept of openness. Translated from the world of software (but not reducible to it), openness must therefore be understood as a powerful new form of political desire in network cultures.[23]

This new stated commitment to the open is not limited to explicitly activist and marginal network cultures. Two radically different but equally "mainstream" organizations, Wikipedia and Google, also understand their operation in terms of openness. As the "free encyclopedia that anyone can edit," Wikipedia is commonly held up as the most successful example of translated openness (which is why I come to focus on it in the following chapters). On the "About" page of the English Wikipedia, the encyclopedia is described as "open to a large contributor base," "written by open and transparent consensus," and the various effects of its "radical openness" are considered.[24] Moreover, the project is built on wiki software, which allows for easy and immediate page creation and modification and is licensed under permissive, commons-based licenses (Creative Commons Attribution–ShareAlike 3.0 License and the GNU Free Documentation License).

Interestingly, while Wikipedia is more often celebrated as the open ideal in terms of contributions, governance, technology, and licensing, the rhetoric of openness is stronger with Google. For example, on Google's Public Policy Blog, Senior Vice President of Product Management Jonathan Rosenberg published a post titled "The Meaning of Open." He writes: "In an open system, a competitive advantage doesn't derive from locking in customers, but

---

23. I use the term "network cultures" to refer to groups who organize, communicate, and produce primarily through the web and other networked technologies.

24. *Wikipedia*, s.v., "About," accessed August 16, 2010, http://en.wikipedia.org/wiki/Wikipedia:About.

rather from understanding the fast-moving system better than anyone else and using that knowledge to generate better, more innovative products."[25] The sentiments expressed in the post are very similar to the ones offered by Raymond and the Open Source Initiative, where openness is figured as an innovative and competitive production method perfectly compatible with new forms of capitalist accumulation. Rosenberg goes on to define openness in terms of hardware and software, information, transparency, and control. The key passage, however, comes toward the end of the post, where he is explicit about what he sees is at stake in the battle for openness:

> Open will win. It will win on the Internet and will then cascade across many walks of life: The future of government is transparency. The future of commerce is information symmetry. The future of culture is freedom. The future of science and medicine is collaboration. The future of entertainment is participation. Each of these futures depends on an open Internet.[26]

For Rosenberg, openness is also a quality of a system. Through competition, the most superior knowledge within that system will rise to the top and continue the march of the system's progress—"better, more innovative products." In a statement verging on both techno-determinism and techno-utopianism, the battle for an open Internet, based on open standards and *system-knowledge competition*, is thought to have flow-on effects that will radically transform all society. The reason for this "cascade" of effects, however, is that in the eyes of Rosenberg the Internet is inseparable from all these "walks of life." Recalling Lessig, the battle for openness is not only about the Internet, but about all society.

I want to finish my review of contemporary openness by considering its deployment outside software, outside network cultures, into the realm of institutional politics and related writings. One of the first to translate the open (back) into institutional politics was Douglas Rushkoff. In 2003 he wrote a short monograph titled *Open Source Democracy: How Online Communication Is Changing Offline Politics*. Rushkoff argues that a new "electronic renaissance" has taken place, a profound shift in the individual's perception of his or her agency in electronic environments and he uses open source as his guiding example: "like literacy, the open-source ethos and process are hard if not impossible to control once they are unleashed. Once people are invited to participate in, say, the coding of a software program, they begin to question just how much of the rest of the world is open for discussion." Rushkoff's

---

25. J. Rosenberg, "The Meaning of Open," 2009, http://googlepublicpolicy.blogspot.com/2009/12/meaning-of-open.html.

26. Ibid.

renaissance, however, does not merely detail how politics can benefit from the insights of open source; it is a politics totally enmeshed in computational metaphors: "The implementation of an open source democracy will require us to dig deep into the very code of our legislative processes, and then rebirth it in the new context of our networked reality" (2003, 56–57).

Less than a decade later, Rushkoff's ideas are fast becoming the norm. For example, the edited collection *Open Government: Collaboration, Transparency, and Participation in Practice* (Lathrop and Ruma 2010) includes contributions by key members of government and commerce and is clearly aimed at a broad audience. The collection's rhetoric is a perfect mashup of Hayek, Popper, Raymond, and Stallman, evidenced by a scan of the section titles such as "Competition Is Critical to Any Ecosystem," "Open Standards Spark Innovation and Growth," "The Closed Model of Decision Making," and "Open Government and Open Society." As mentioned earlier, the connection to the organizational method of software cultures is made more explicit in Tim O'Reilly's contribution, "Government as a Platform," where he writes:

> What if . . . we thought of government as the manager of a marketplace? In *The Cathedral & the Bazaar*, Eric Raymond uses the image of a bazaar to contrast the collaborative development model of open source software with traditional software development, but the analogy is equally applicable to government. . . . A bazaar . . . is a place where the community itself exchanges goods and services. (O'Reilly 2010, 11)

The rest of the piece is dedicated to translating Raymond's insights to the practice of government. In another chapter, Charles Armstrong describes the profound impact the Internet holds for democracy:

> The Internet has changed a fundamental aspect of democratic systems which has persisted for 7,000 years. The change may presage a period of democratic innovation on a scale comparable to classical Greece. It will lead to democratic systems that are more fluid, less centralized, and more responsive than those we know today; systems where people can participate as little or as much as they wish and where representation is based on personal trust networks rather than abstract party affiliations. This is Emergent Democracy. (Armstrong 2010, 167)

Upon considering "the road to emergent democracy," Armstrong notes that "we tend to associate democracy with nations, cities, and other state entities," but it is his "hunch that virtual corporations" and not traditional institutions will pave the way (175). The example of what these "virtual institutions" might look like, surprise surprise, is Wikipedia. In both of these texts government is reimagined as a competitive marketplace of ideas, modeled after bazaar-like "virtual corporations" that resemble Wikipedia and that promise to reinvigorate democracy on a scale unmatched since classical Greece.

This kind of sentiment is similarly expressed by serving politicians and newly established government initiatives, such as Obama's Open Government Initiative that I detailed in the opening lines of the introduction. One early supporter of the open in Britain was the conservative MP Douglas Carswell. In 2009 Carswell appeared on the current affairs show *Newsnight* specifically in relation to the parliamentary expenses scandal, which later led to the stepping-down of the Speaker of the House of Commons. After commenting critically on the scandal, Carswell called for a move to "Open Source Politics." This gesture was mirrored on his blog, where he writes: "Open source software. Wikipedia and wiki-learning. Open source parties and politics, too?" He also writes about opening the primaries (the selection of candidates for election)—"Open primaries might spell the end for closed-shop parties"—and name-drops Clay Shirky and notions like "collaborative creation."[27]

Australia's Government 2.0 Taskforce represents a more organized statement of political openness. On the "About" section of its website the taskforce divides its stated areas of activity into two streams:

> The first relates to increasing the openness of government through making public sector information more widely available to promote transparency, innovation and value adding to government information.
>
> The second stream is concerned with encouraging online engagement with the aim of drawing in the information, knowledge, perspectives, resources and even, where possible, the active collaboration of anyone wishing to contribute to public life.[28]

Finally, the conservative libertarian movement, the Tea Party, has also borrowed ideals from open source. The production of the movement's notorious "Contract from America," a document that lists ten key agenda items the movement would like congressional candidates to sign, was described by its creator in the *New York Times* as follows: "Hundreds of thousands of people voted for their favorite principles online to create the Contract as an open-sourced platform for the Tea Party movement" (Becker 2010). Openness, information, collaboration, transparency, and participation—government 2.0 indeed.

---

27. D. Carswell, "Open the Party; Open Source Politics Is Coming," Douglas' Blog, 2009, http://www.talkcarswell.com/show.aspx?id=926.

28. "About," Government 2.0 Taskforce, 2010, http://gov2.net.au/about.

Thus far I have traced the open through debates about computer systems and standards, into software cultures in the eighties and nineties, network cultures over the last decade, and finally outside of these realms and into institutional politics and society more generally. And while the notion of openness in government is not unique to the current neoliberal regime (as the writings of Popper make clear), I hope to have made clear that the general deployment of the open in institutional politics, and as a political concept more generally, cannot be separated from its emergence in software and network cultures. Indeed, it is perhaps more accurate to posit that today's openness is evidence of the networked and computational, even cybernetic, nature of governance. Through these multiple trajectories, openness is placed in a variety of settings, articulated alongside different concepts, and put to use in different ways. The open circulates, scales up, garners new allies, is reconfigured, distinguished, and remixed; each movement troubles and destabilizes the articulation of its meaning. The open sways between means and ends, between noun, verb, and adjective. And throughout all this movement, openness nonetheless maintains certain consistencies, such as its couplings with transparency, collaboration, competition, and participation and its close ties with various enactments of liberalism. What to make of a concept championed by all walks of political life? When conservative liberals, libertarians, liberal democrats, postautonomous Marxists, and left-leaning activists all claim the open as their own and all agree that openness is the way forward? What to make of a politics that contains bits and pieces of many older political positions, but cannot be aligned easily with any one in particular? Or can effortlessly be deployed on any scale, from small projects to entire societies, by the American government or the more radical P2P Foundation? One that defends markets but (at least in some instances) attacks (intellectual) property, and whose meaning is so overwhelmingly positive it seems impossible even to question, let alone critique? To be sure, we can attack something for not really being open, for not living up to the ideals of openness, but can we take aim at openness itself?

While the *force of the open* must be acknowledged—the real energy of the people who rally behind it, the way innumerable processes have been transformed in its name, the new projects and positive outcomes it has produced—I suggest that the concept itself has some crucial problems. In the final section of this chapter I aim to demonstrate that not only is the open problematic in relation to contemporary software and network cultures, but that the concept contains a poverty that has existed in all its uses throughout history and that makes it unsuitable for political description. Indeed, I argue that the open actively works against the development of a political language—if, that is, we take the political to extend beyond questions of just governance to the circulation and distribution of power and force, and take politics to mean the distributions of agency in general as well as the conflicts and issues that emerge when antagonistic flows intersect. To make this argument, I return to Popper and *The Open Society*.

A Critique of Political Openness

The various criticisms Popper levels against Plato and his followers have not gone unchallenged (Cornforth 1968; De Cock and Bohm 2007; Kendall 1960; Levinson 1970; Shearmur 1996; Vernon 1976). The critique I put forward does not rely on whether Popper was faithful in his interpretation of the great philosophers, or of the terrible events that inspired his texts, as do others. I do not interrogate how Popper's political philosophy is intimately connected if not derived from his scientific philosophy and I do not challenge the validity of the unique concepts he relies on in *The Open Society*, such as "critical rationalism" and "piecemeal social engineering" (see Adorno et al. 1976). Instead, I focus specifically on the character of openness per se in relation to the rest of his text. I argue that the logic of openness actually gives rise to and is perfectly compatible with new forms of closure; indeed, that closure is inherent in Popper's notion of openness. Moreover, I argue that there is something about openness, about the mobilization of the open and its conceptual allies, that actively works against making these closures visible. I finish by reflecting on the peculiar situation of the second coming of openness within the supposedly already-open society.

Like his critique of the closed—which has both conceptual and empirical dimensions—Popper's open society is both a set of ideas and a really existing entity. These two dimensions, which enact an awkward distinction between reality and thought, play out and interact in complex and ultimately troubled ways. The first thing one notices about *The Open Society* is that it is almost entirely dedicated to critique. In the first volume, for example, the open is not taken up at all until the final chapter and only after nine preceding chapters on Plato. It is a work explicitly concerned with the *Enemies* of the open, rather than the open itself (Magee 1982, 87). As a concept, therefore, the open is reactionary; it gains meaning largely through a consideration of what it is *not*. The open is significant in terms of the actually existing political situation of his time, for example, largely because it is neither fascism nor communism. The alternative, liberal democracies with muzzled forms of capitalism, which represent the modern incarnation of the open society, seemed preferable to these other political programs. As we shall see, this negative, or *is not*, quality of the concept and the concomitant reluctance to build a lasting affirmative dimension is entirely necessary and gets to the heart of the problem of the open as a political concept.

Of the negative or *is not* qualities of the open, we can extract the following from Popper's critique of closed societies: open societies do not condone historical or economic determinism, do not support programs of radical social engineering based on truth claims, and do not hold any truth to be absolute.[29] These qualities are the direct result of the critique of the closed I detailed earlier. Under closer analysis, however, it is possible to identify numerous positive qualities of open societies, even though these are generally mentioned in passing and without extended elaboration in the text. Such positive qualities surface especially in the final chapter of the first volume of *The Open Society*, after the critique of Plato is concluded. In these pages Popper regularly invokes reason and the rational as characteristics of open societies, noting for example that the open is "a rational attempt to improve social conditions" (1962, 1:172).[30]

Open societies are also individual-centric in terms of decisionmaking, responsibility, competition, and familial ties. According to Popper, "Personal relationships of a new kind can arise where they can be freely entered into, instead of being determined by the accidents of birth; and with this, a new individualism arises" (175), and "the society in which individuals are confronted with personal decisions, [is] the *open society*" (173). Such individualism also leads to strong competition and exchange relations: "one of the most important characteristics of the open society [is] competition for status among its members," and further, "in an open society, many members strive to rise socially, and to take the place of other members. This may lead, for example, to such an important social phenomenon as class struggle" (174). Regarding exchange, Popper writes, "our modern open societies function largely by way of abstract relations, such as exchange or co-operation" (175). While discussing the historical beginnings of the open society, Popper offers the following quasi-legal and ethical characteristics: "The new faith of the open society, the faith in man, in equalitarian justice, and in human reason, was perhaps beginning to take shape, but it was not yet formulated" (1962, 189). Further-

---

29. He writes, "But we should be clear that there cannot be any theory or hypothesis which is not, in this sense, a working hypothesis, and does not remain one. For no theory is final, and every theory helps us to select and order facts" (Popper 1962, 2:260).

30. Further along he writes, "Once we begin to rely upon our reason, and to use our powers of criticism, once we feel the call of personal responsibilities, and with it, the responsibility of helping to advance knowledge, we cannot return to a state of implicit submission to tribal magic. For those who have eaten of the tree of knowledge, paradise is lost" (ibid., 1:200).

more, he states, "Individualism, equalitarianism, faith in reason and love of freedom were new, powerful, and, from the point of view of the enemies of the open society, dangerous sentiments that had to be fought" (1962, 199). Finally, throughout the text as a whole Popper regularly makes gestures to democracy as preferable to totalitarian regimes. Indeed, at times the open society appears interchangeable with Popper's understanding of democracy.

Of these negative and positive identifiers, it is only the negative qualities that approach anything like the essence or definitive core of openness. More precisely, the positive qualities of openness are actually negative qualities masked as positive ones, or alternatively exist at the level of *reality* (of real practices) and are therefore subject to continual transformation. Openness emerges as a theory bereft of content coupled with a really existing practice, defined by its continual nonidentification with itself. This character of openness is made clear through a consideration of Popper's understanding of democracy—one of the key *positive* qualities of the open society. He writes:

> The theory I have in mind is one which does not proceed, as it were, from a doctrine of the intrinsic goodness or righteousness of a majority rule, but rather from the baseness of tyranny; or more precisely, it rests upon the decision, or upon the adoption of the proposal, to avoid and resist tyranny. (1962, 124)

Popper proceeds immediately to distinguish between two "types of government": one that can be removed "without bloodshed," through institutional processes such as elections, and the other, which can only be removed via revolution. Only the first mode of governance is democratic: "I suggest the term 'democracy' as a short-hand label for a government of the first type, and the term 'tyranny' or 'dictatorship' for the second." So, a "theory of democracy" on the one hand, a theory defined as against tyranny, or *not tyranny*, and a practice of democracy on the other, with institutions and processes and all the messy details that practice implies.

I want to suggest at this point that all of these positive, if ephemeral, traits of the open are alive and well in contemporary society. They remain, in fact, the central values of neoliberal democracies: freedom, democracy, individualism, competition and exchange (free markets), equalitarian justice, and reason. This is hardly surprising, considering that Popper's (abstract) concept of the open society in fact corresponds with the actually existing capitalist democracies of the mid-twentieth century and such societies persist (this is not to suggest that such characteristics are indisputably realized or haven't changed their meaning and function, but rather that politics still plays out in and through this language). It seems, therefore, that we are still in the society

Popper described as open, both in terms of *is not communism or fascism*, and in terms of the positive qualities just mentioned. At the same time, however, and consistent with Popper's logic, this actually existing open society is likely to change.

To continue with the account of democratic practices, for example, it is likely and indeed categorically necessary that these practices may be replaced. How is it that *specific sets of practices* called "democracy" are part of the open and yet in future might not be? One possible response is that the democratic practices might be succeeded by something that is even more democratic and thus even more open. Another possible response is that these practices have *become* closed, that somehow, through time, this mode of governance loses it character of openness. Both of these responses, however, suggest that *forms of closure* exist *within* open societies: in the first scenario, if a practice is able to be replaced with a *more open* one, it must not be entirely open to begin with. The second scenario indicates that the seeds of closure are already immanent within this open mode of governance. However, the alternative, which is to affirm a specific version of democracy and a specific program of knowledge and related practices—in short, a precise truth of the open—is simultaneously the open's closure. Thus, by invoking positive but ephemeral qualities, and a society that necessarily changes, Popper avoids the kinds of closure he identifies in totalitarian thought. At the same time, closure remains an inherent part of the open; it is what openness must continually respond to and work against—a continual threat among the ranks.

Openness, we might say, implies antagonism, or what the language of openness would describe as closures. Such closures do not randomly emerge, unexpected and from the outside. It is the very qualities that Popper holds up as representative of contemporary openness and that constitute the formal language of the just organization of society—freedom, competition, equality, and exchange, coupled with democratic institutions—that not only coincide with but are actually generative of new forms of closure. The most obvious of these are the economic closures produced by "competition" and "free markets," of which the 2008 global financial crisis is only the most recent and dramatic episode of the continuous and generalized asymmetrical distributions of agency produced by (debt-based) informational capitalism. These are the same general conditions—call them conditions of *im*possibility—that produced the invisible source code, nondisclosure agreement, and broader regime of intellectual property that Stallman experienced as closures in his lab at MIT. In short, Popper's argument against totalitarian knowledge is compatible with, and even constitutive of, neoliberal capitalism. And it is these same forms of closure that the second coming of openness, together with its new set of conceptual allies, tries to address. But what to make of this second coming?

The first thing one notices is the curious situation of openness emerging within a supposedly already-open society. Other than confirming the closures inherent in the open, I think this curious situation is suggestive of a more crucial conceptual shortcoming of openness. Once a form of organization is labeled open, be it a state, movement, project, or party, it becomes difficult to account for the politics (closures) that emerge from within. For Popper, this is because his version of the open is primarily a critique of totalitarian knowledge, but also because he struggles to focus on the details of his open society for fear of closing it. Recent uses of openness—from open systems, to open source, and free and open everything—bear significant resemblance to Popper's in terms of character and function. Once more the open emerges largely as a reaction to a set of undesirable developments, beginning with the realm of closed systems and intellectual property and its "closed source." And once more the open is articulated alongside an entourage of fractal subconcepts that defer political description, of which I will soon come to focus on participation, collaboration, ad-hocracy, and forking. While this re-emergence works as a critique of Popper-Hayek openness, it simultaneously reinstates the same conceptual architecture. Of all the authors cited in the account of openness I have developed here, for example, very few have turned a critical eye to the open, and there has been very little criticism about specific open projects. If a critical word is written, it is rarely substantial and most likely about how one small component can be made better, more open.[31] Somewhat ironically, once something is labeled open, it seems that no more description is needed. Recalling Kelty's remarks, openness is the answer to everything and it is what we all agree upon.

I began this chapter by quoting radically different thinkers, Lessig and Hardt and Negri, who all gestured to the open. Throughout my analysis I added many names and organizations to these three, from the Tea Party to

---

31. One notable exception is an essay by Jamie King titled "Openness and Its Discontents" (2006). King makes similar observations about the rise of openness—"openness is clearly becoming a constitutive organizing principle"—and its political blind spot—"if we are going to promote open organization . . . we must scrutinize the tacit flows of power that underlie and undercut it." He finishes with this insightful observation regarding the logical flaw of absolute openness: "the most open system theoretically available reveals perfectly the predicating inequalities of the wider environment in which it is situated" (2006, 45, 53). More recently Evgeny Morozov has written a chapter on openness and transparency in *To Save Everything, Click Here* (2013), although Morozov's focus is much more on the negative effects of releasing government data.

"leftist" activist groups, governments, major corporations, and scholars. All these individuals and groups understand their practices and ideas in relation to the open and use it to "look forward." Despite any differences, there is clearly a coherent language for supporters of the open and an array of subconcepts that help distinguish openness from other political modes of existence.

I hope to have shown, however, that historically the open has not proven well suited to this task of looking forward. Rather than using the open to look forward, there is a pressing need to look more closely at the specific projects that operate under its name—at their details, emergent relations, consistencies, modes of organizing and stabilizing, points of difference, and forms of exclusion and inclusion. If we wish to understand the divergent political realities of things described as open and to make visible their distributions of agency and organizing forces, we cannot "go native," as the anthropologists might say, meaning we cannot embrace the frameworks of understanding used in the practices we wish to study. To describe the political organization of all things open requires leaving the rhetoric of the open behind.

A Note on Method

At this point, the question of how to proceed, of how to describe the political conditions of openness without relying on its language, is pressing. As I stated earlier, the following chapters all consider different aspects of the most successful open project to date, Wikipedia. My focus is on this single project and on specific moments or scenes within this project that have a certain political and organizational pertinence. These organizational scenes all take place within the paradigm of openness, each relating to, but also disturbing, one of its subconcepts (participation, collaboration, ad-hocracy, and forking). But how should this be done? Which methodological framework, each with its own distinct ontological and epistemological presumptions, is fitting? These questions form an ongoing theme and I do not attempt to address them comprehensively here. However, some preliminary observations on method—to be revised, fleshed out, and extended—are required.

Large parts of the following chapters are concerned with the analysis of *statements*. The notion of "statements" is indebted to Michel Foucault's method of archaeological discourse analysis, where the individual statement forms the "atom of a discourse" (1972, 90). Foucault develops his notion of statements by distinguishing them from other well-known and at times overlapping notions: sentences (from linguistics), propositions (from logic), and speech acts (from pragmatics). To simplify slightly, statements can be understood as the "raw material" that may, upon satisfying further criteria, later be identified as a sentence, proposition, speech act, or something else entirely, but whose primary domain of existence is historical and ontological (rather than, e.g., formal, logical, or grammatical). And while all statements exist on a spectrum of performativity, while they all enact and contribute to different realities, this performative dimension is more general, more difficult to ascertain, and more material than the speech acts first described by Austin (more on this later).

Departing slightly from Foucault, I understand statements as corresponding to the generic notion of *inscription*: anything that *inscribes the world* produces statements. Forms of inscription include not only traditional writing and speech (Derrida 1997; Ong 2002) but also the various media attended to by media theorists and archaeologists (Kittler 1990, 1997, 1999, 2010; Zielinski 2006; Huhtamo and Parikka 2011; Parikka 2012; Ernst 2012), as well as the scientific "inscription machines" attended to by those in science and technology studies (Latour and Woolgar 1986; Mol 2002; Keating and Cambrosio 2003).[32] There are, in other words, innumerable types of statements, with different material existences, varying durabilities, belonging to distinct ordering principles, positioned specifically within a network of related statements and in distinction to others. As Foucault puts it:

> At the very outset, from the very root, the statement is divided up into an enunciative field in which it has a place and a status, which arranges for its possible relations with the past and which opens up for it a possible future. Every statement is specified in this way: there is no statement in general, no free, neutral, independent statement; but a statement always belongs to a series or a whole, always plays a role among other statements, deriving support from them and distinguishing itself from them: it is always part of a network of statements, in which it has a role, however minimal it may be, to play. (1972, 111)

An analysis of statements is therefore not an exercise in hermeneutics—it does not concern itself with the meaning of an encyclopedia article or a particular policy, for example—and it is not interested in whether or not a particular proposition is true (although it is interested in the general conditions of truth). As Deleuze and Guattari might put it, when statements are manifested as language, the question is not what does it mean or is it true, but what does it do? Their answer would of course be "statements order" and thus an analysis of statements is necessarily concerned with the questions of order

---

32. Two authors who have specifically sought to introduce the notion of statements into computational media are Parikka (2007) on computer viruses and Goffey (2008) on algorithms.

and organization.³³ But it is also to inquire more specifically as to how statements relate to one another; what makes some statements possible (and others impossible) within a formation; how are statements arranged and internally differentiated within a formation and what status are they accorded?

As is well known, the political dimension of Foucault's method is made explicit in his coupling of discourse (knowledge or truth) with power, where he writes, for example, that "there can be no possible exercise of power without a certain economy of discourses of truth which operate through and on the basis of this association" (1980, 93). My own account of statements similarly emphasizes its political dimensions, although in ways that differ somewhat from Foucault. The most obvious point of difference is the smaller scale of my analysis and its stronger contemporary focus. Rather than mapping an entire discourse or tracing a long genealogy (say, of encyclopedic knowledge), I focus on the arrangement of statements that mark the limits of a single project. I call the totality of statements that constitute a project like Wikipedia a *statement formation*. Statement formations are coherent and ordered, but they do not constitute an entire discourse. They are empirical aggregations, not necessarily or not only brought together by the rules of a discourse, but by any number of forces; they may be strongly ordered by one discourse in certain sections—such as the discourse of encyclopedic knowledge—but they also consist of other statements organized by different rules, such as rules of etiquette when discussing articles or the rules of a programming language when modifying the wiki software.³⁴ If Foucault's archaeology and genealogy extended outward across different knowledge spaces or through different temporalities, statement formations are sites in the present (with their own histories, to be sure), where statements possibly belonging to quite different orders of discourse encounter one another.

Alongside current understandings of Wikipedia as an "open project," therefore—where everyone can participate and working together takes the form of collaboration; where governance is ephemeral and ad-hocratic; and where leaving takes place through the virtually lossless procedure of forking—I approach Wikipedia as a statement formation. This shift, which forms the basis of my attempt to build a politics without recourse to openness, re-

---

33. For Deleuze and Guattari's related discussion of "order words" see Deleuze and Guattari 1988, 83–122.

34. I see this smaller scale and more empirical style as broadly consistent with insights put forward by John Law in *After Method* where he explains the differences between Foucault's method with that of Latour and Woolgar (Law 2004, 35–36).

frames the working together of collaboration as a question of how new statements enter a formation; the ad-hocratic governance of Wikipedia as a question of how to make visible the project's most forceful statements, the ones that *organize* the formation and make it durable; and in place of the lossless exit of forking, I explore, among other things, what happens when powerful conflicting statements are forced together within a formation.

2

Sorting Collaboration Out

> We're really just at the beginning, still, of collaborative efforts.
> JIMMY WALES (2008)

Participation and Collaboration

Without getting into dead-end arguments about determinisms, it has been clear at least since the pioneering work of Harold Innis (1950, 1951) that the characteristics of a medium bear considerably on the shape of the collectives that mobilize through and in relation to it. It is, in fact, misleading from the very beginning to consider the two as separate or separable. Television, for example, configured the social in very specific ways, constituting some as part of a new kind of individualized but "mass" audience (Williams 1976, 196), as primarily receivers of content, and a much smaller group as producers and distributors of content, sometimes critically referred to as "elites." While Raymond Williams famously noted that there are no actual masses, only *ways of seeing* people as masses, it is perhaps more accurate to note that there are no pre-existing masses, only technological configurations that constitute people as masses. So-called one-to-many broadcast models of media mirrored the dominant industrial models of the second half of the twentieth century and their tendency to massify. Of course, such configurations are never totalizing or exhaustive, and critical perspectives based on generalities of form rarely capture the complexity of relations that take place within and around a given technical artifact. A key historical example in relation to television is Henry Jenkins's work on "textual poachers," active audiences who complicate notions of passivity and, indeed, of the mass (1992). A key theoretical influence on Jenkins's text was Michel de Certeau's brilliant book *The Practice of Everyday Life* (1984), which was at least in part a response to

---

Epigraph: "Internet Collaboration Still in Infancy," *Sydney Morning Herald*, November 3, 2008, http://www.smh.com.au/news/technology/web/internet-collaboration-still-in-infancy-wikipedia-founder/2008/11/03/1225560695451.html.

the disciplinary "microphysics of power" developed by Foucault (1977). If Foucault already noted that power is in the details, de Certeau was equally trying to uncover the "popular procedures (also "miniscule" and quotidian) [that] manipulate the mechanisms of discipline and conform to them only in order to evade them" (1984, xiv). Jenkins's work is a sober corrective to the persistent trope that old broadcast media create passive audiences and new media active users. Relations of agency are and have always been more complicated.

The reason for this brief historical vignette, though, is that Jenkins was one of the first to try to come to terms with transformations in production brought about by the distributed, many-to-many architectures of the Internet. In *Textual Poachers*, Jenkins had developed the notion of "participatory culture," groups of fans who actively contribute in various ways to popular texts and narratives. This notion of participatory culture was mobilized and refined in relation to networked mediations of production (2006a, 2006b, 2009). For Jenkins, the practices characteristic of fans—whose active response to broadcast included not only critical and contested readings, but also the production of related amateur content—become a starting point for addressing the new models of production emerging on the web. He defines such media cultures as those "with relatively low barriers to artistic expression and civic engagement, strong support for creating and sharing creations, and some type of informal mentorship whereby experienced participants pass along knowledge to novices" (2009, xi). Jenkins describes not so much the specificities of working together in networked ecologies as the wider conditions in which new forms of working together take place: low barriers to entry, support for creating and sharing, and informal mentorship.

Jenkins's observations of a participatory turn were paralleled by other observers in adjacent areas of study, such as Clay Shirky's *Here Comes Everybody* (2008), Howard Rheingold's *Smart Mobs* (2002), and Yochai Benkler's notion of "commons-based peer production" developed in *The Wealth of Networks* (2006). Ideas of the mass audience and its retrospectively strengthened connotations of passivity are replaced by ones of intelligent bodies of peers in motion, whose relationship to production is both direct and multiple. The mere fact of participation, however, doesn't provide substantial inroads into understanding transformations in actually working together—the qualitative or micro-dimensions. And thus, within this paradigm of participation it is the term *collaboration* that has been put forward to do this more specific work. Noncoincidentally, a renewed interest in collaboration emerged almost exactly with the rise of Wikipedia as a popular artifact. Formations like Wikipedia pose the problem of working together that collaboration, as a specific type of working together, attempts to solve (Benkler 2006; Bruns 2008a; Elliott 2006; Jenkins 2006a; Reagle 2010; Shirky 2008).

For the legal scholar Yochai Benkler, collaboration forms one of the core features of his "commons-based peer production." He writes:

> The networked environment makes possible a new modality of organizing production: radically decentralized, collaborative, and non-proprietary; based on sharing resources and outputs among widely distributed, loosely connected individuals who cooperate without relying on market signals or managerial commands. (2006, 60)

Collaboration is a kind of working together that isn't directed by economic imperatives or top-down orders; it exists outside the rules of commodity exchange and the hierarchies of firms and governments. Benkler's notion of collaboration being "radically decentralized" refers not only to geographic or spatial qualities, but equally or even primarily to notions of autonomy within forms of organization. Benkler writes that "individual action is self-selected" and "the actions of many agents cohere and are effective despite the fact that they do not rely on reducing the number of people whose will counts to direct effective action" (62), which is of course what centralized, hierarchical modes of organizing do. An explicit link is made between decentralization and the ideal market, with the difference being that commons-based peer production does not rely on the price mechanism. The parallels between Benkler and the logic of openness I identified in Popper, Hayek, Raymond, and others are quite visible. Even more so given the fact that Benkler holds up an idealized notion of science very similar to Popper's—and probably quite unrecognizable to the average scientist—as a historical instance of his vision of collaboration:

> This kind of information production by agents operating on a decentralized, nonproprietary model is not completely new. Science is built by many people contributing incrementally—not operating on market signals, not being handed their research marching orders by a boss—independently deciding what to research, bringing their collaboration together, and creating science. (Benkler 2006, 63)

Besides being radically decentralized and nonproprietary, there are a whole host of other characteristics and processes typical of this new collaborative work put forward by Benkler and his counterparts. In line with the participatory turn, collaborative work begins with default openness. Benkler refers to Wikipedia's "anyone can edit" slogan, while the creative industries scholar

Axel Bruns writes of "open participation" (2008a, 24). Similar to the version-labeled products of software (1.0, 1.1, 1.2, and so on), the outputs of collaborative work are never finished in the sense that is true of most industrial goods. Bruns describes this mode of production—or his updated notion of "produsage"—as a "continuing process" whose outputs are "unfinished artefacts" (2008a, 27). Likewise, Clay Shirky writes, "A Wikipedia article is a process, not a product, and as a result, it is never finished" (2008, 119). Collaboration is also thought to be most successful when the work at hand can be broken down into small, discrete or semi-discrete components. Benkler refers to a project's "modularity," the extent to which it can be broken down into parts, and of "granularity," which refers to the size of a module in terms of the time and effort taken to complete a task (2006, 100). Wikipedia would be considered highly modular and granular because tasks can be broken down into very small components. As Shirky puts it: "Some [contributors] add new text, some edit the existing article, and some fix typos and grammatical errors" (2008, 118).

Within this unending modular work, Shirky suggests that there is an "unmanaged," "spontaneous division of labor" (2008, 118), similar to what Axel Bruns calls "ad hoc meritocracy" (2008a, 25). The nature of a contribution, of who does what, is described as the spontaneous outcome of the current state of the article (if the article exists already) combined with the skill set and level of commitment of the person contributing. This leads to scenarios where "one person can write a new text on asphalt, fix misspellings in Pluto, and add external references for Wittgenstein in a single day" (Shirky 2008, 120). While Shirky rightfully points to the novelties of this way of working, his "spontaneous division of labor" overlooks the many planned divisions of labor, such as the formation of WikiProjects and task forces that focus on specific topics or particular aspects within a given topic. There are also other more obvious divisions of labor that emerge around the permissions-based hierarchy of editors. That is, as contributors climb the ranks of Wikipedia, they are granted new capacities such as the ability to delete articles, block other people from editing, make edits to "protected" articles, and so on (see chapter 3). Finally, while contributions might seem spontaneous when observing the archived list of edits (Wikipedia page history) or in comparison to a Fordist production line, it is less so if the perspective switches to the contributor (e.g., their knowledge base or skill set) or the actual article and its current state.

Another writer to tackle the division of labor and the organization of collaborative work more generally is Mark Elliott. Rather than stress the random nature of the distribution of tasks, Elliott describes them as "stigmergic." His term "stigmergic collaboration" is borrowed from entomology (the study of insects):

> a particular configuration of a termite's environment (as in the case of building and maintaining a nest) trigger[s] a response in a termite to modify its environment, with the resulting modification in turn stimulating the response of the original or a second worker to further transform its environment. Thus the regulation and coordination of the building and maintaining of a nest [is] dependent upon stimulation provided by the nest, as opposed to an inherent knowledge of nest building on the individual termite's part. A highly complex nest simply self-organizes due to the collective input of large numbers of individual termites performing extraordinarily simple actions in response to their local environment. (Elliott 2006, para. 3)

The idea of stigmergy attempts to solve the puzzle of mass collaboration: how thousands of people can organize outside of the managerial structures or price mechanisms described above by Benkler. With stigmergic collaboration, contributions themselves are *a form of communication about the larger structure* as well as the work that remains to be done, even though no individual has command over the larger structure. In this sense, there is an odd situation where the "modular," "granular" tasks somehow produce a coherent whole all on their own. In Elliott's language, it is almost as if stigmergy enables bottom-up cathedrals, rather than the bazaars described by Eric Raymond (2000). While I think there are important limitations in translating the idea of stigmergy from its insect origins, most notably that it risks equating the capabilities of humans with those of ants (see Tkacz 2010), it is nonetheless interesting in that it suggests a peculiar form of distributed agency in relation to organization: there is something in the contribution itself—whether in the actual act of contributing or the result—that *communicates organization.* One contribution "triggers" others and also acts a model for them. Later I will stress this inherent notion of organization with regard to Gregory Batson's notion of framing and metacommunication (which is different than self-organization).

If collaboration is "open to anyone," if managerial hierarchies are replaced with peers, and if there is no traditional market (or price mechanism) to organize value, how does collaboration sort desirable contributions and contributors from undesirable ones? How does one judge what is good, what belongs, and what doesn't? These are questions of organization, order, stability, sorting, and in some instances governance, and will occupy much of the next two chapters, where I argue that there is no organization, no coherent thing, without some form of differentiation from "what it is not" and without mechanisms that sort and filter what is part of the thing and what is not. How has this question of organizing, of creating order and sorting good from bad, been handled within the paradigm of collaboration?

It is important to note from the outset that these kinds of questions play out differently in different projects. Benkler, for example, uses a series of examples (Google, Amazon, Slashdot, the Open Directory Project) and considers these issues of "relevance" and "accreditation." He poses the question thus: "How are we to know that the content produced by widely dispersed individuals is not sheer gobbledygook?" (2006, 75). All the projects Benkler mentions have different ways of determining and ranking relevant contributions. Google, for example, uses an algorithm that maps how many incoming links a site has (how many other sites links to it), to determine how high a website appears on search results. While Google's search algorithm is not an open project, it nonetheless "harnesses" the linking activities of distributed users to produce a relevance index. "Geek" news site Slashdot has a system of rating the comments section of news stories. Contributors build up "karma" depending on how their comments are rated. There are also rules about who can rate comments, such as those doing the rating must themselves have "positive karma," and the number of ratings allowed per person is limited to ensure that no single contributor has too much sway over which comments rise to the top and which ones remain relatively invisible (2006, 78). This process determines what is relevant and sorts contributions accordingly. It is a system of ranking contributions that in turn ranks contributors.

Axel Bruns describes very similar processes in a lengthy discussion of Wikipedia, but he develops concepts that focus more on the sorting of people rather than contributions. His notion of "produsage," referred to above, tries to capture certain forms of "collaborative content creation" where the moments of production and consumption become difficult to separate (2008a, 24). More specifically, produsage refers to activities where the very act of consumption is also an act of production (such as with Google Search, where the act of searching produces data that is fed back into the algorithm, effectively contributing to the Google Search "product"). Along with the "open participation" mentioned above, Bruns outlines three other qualities of produsage that are relevant: communal evaluation, fluid heterarchy, and ad hoc meritocracy (2008a, 24–26). Communal evaluation occurs when contributions are evaluated by the other contributors. Bruns notes that this often happens implicitly, with useful contributions being taken up and improved upon and poor contributions being ignored. Bruns also suggests that this form of evaluation has a "socializing" effect on participants:

> This holoptic model of communal evaluation in produsage, in which each contributor is able to see and evaluate everyone else's contributions, also acts as a driver for a continuing process of socialization of participants into the community ethos: being able to view all of their peers' contributions provides individual members with a clear understanding of the forms and formats their own contributions may take, and the quality and quantity of input required if they wish to become a more central member of the community. (2008a, 25)

Communal evaluation thus generates something akin to norms, which shape what and who are on the inside and the margins, respectively.

Because communal evaluation happens constantly, and thus people and contributions are also evaluated constantly, Bruns argues that there are no hierarchies in collaborative, produsage projects. Rather, and related to Shirky's "spontaneous division of labor," Bruns writes of "fluid heterarchy." Instead of labor itself being "spontaneously organized," it is the people in relation to their labor and each other who organize spontaneously. What makes fluid heterarchy different from "traditional hierarchical organizational models" is that "leaders emerge from the community based on the quality of their contributions" (2008a, 25). It seems, however, that even though "traditional" forms of organization (such as bureaucracies) have their own flaws, many of these forms too are based on the idea that the organization of members is also based "on the quality of their contributions," as Bruns puts it for produsage collaborations. In fact, this is what supposedly defines many modern forms of organization over more traditional ones, such as monarchies. Finally, because these collaborative projects are heterarchical and constantly rearranging people based on the quality of their contributions, Bruns describes them as "*ad hoc* meritocracies" (a notion that I historicize in the following chapter); the heterarchy will change in relation to the merits of the contributors in a somewhat "on the fly" fashion.

While none of these writings on collaboration is reducible to the others, they all attempt to grasp an emergent form of working together, made possible by the distributed architectures of the Internet, together with novel intellectual property licenses that fall outside both centralized forms of production (governments) and production organized around the laws of the capitalist market (although in discussing the work of Raymond I have suggested that market-based thinking is not jettisoned but reappropriated). They also all claim, in different ways, that this new form of working together is nonhierarchical, or, at least that hierarchies are fleeting and *just* because they are based on merit. Collaboration is also consistently depicted as a "looser" form of organization, with individual roles always contingent and shifting.

SORTING COLLABORATION OUT   49

Alongside these theoretical similarities, Bruns, Benkler, and Shirky all note that Wikipedia is brought together by a shared desire to make *an encyclopedia* and as such there are rules that must be followed. Benkler puts it like this: "When [people] enter the common project of *Wikipedia*, they undertake to participate in a particular way—a way that the group has adopted to make its product be an encyclopaedia" (2006, 73). Bruns (2008a) writes that

> a small number of fundamental policies act as the foundational law of the *Wikipedia* project. Their existence should not be misunderstood as an anachronism amid the open and flexible environment of produsage itself—indeed, the core principles of *Wikipedia* crucially serve to define the fundamental purpose of the project itself, and ensure its continued feasibility.

It is the relation—or perhaps tension—between the flat, peer-evaluated, and merit-based forms of ordering people and the many (it is indeed many) rules that constitute the project as a unique encyclopedia that I now wish to explore in more detail. It is a relation that doesn't feature strongly in the existing literature on collaboration, with the exception of Joseph Reagle's discussion of Wikipedia's policies—the neutral point of view (NPOV) in particular—in his chapter on "Good Faith Collaboration" (2010, 45–72).

Through a consideration of the entry on evolution, Reagle shows how collaboration is actually only possible because of NPOV, together with the assumption of "good faith" (2010, 45–55). NPOV (considered in more detail in the following chapter) is the mechanism that seemingly allows both believers and critics of the theory of evolution to work together and, importantly, to work out any conflicts that arise in the process. Reagle also digs up the following quotation from Wales which suggests that Wikipedia's cofounder sees NPOV in a similar way:

> The whole concept of neutral point of view, as I originally envisioned it, was this idea of a social concept, for helping people get along: to avoid or sidestep a lot of philosophical debates. Someone who believes that truth is socially constructed, and somebody who believes that truth is a correspondence to the facts in reality, they can still work together. (Wales, cited in Reagle 2010, 53)

From the perspective of Reagle and Wales, therefore, collaboration is the result of certain principles that seemingly allow everyone to work together regardless of their particular point of view. Wikipedia is collaborative not because it has no hierarchies, but because it has policies that mediate between different and, indeed, often conflicting views, seemingly absorbing different perspectives into a single frame. While Reagle's work is exceptional in the way it couples the concept of collaboration with core Wikipedia policies, the perspective I develop throughout this chapter and the next is less certain about the possibility for NPOV and other core policies to mediate different "points of view" and the disputes that emerge from them.

In what follows, I consider two instances of collaboration or, rather, failed collaboration in Wikipedia—two events that fall under the paradigm of participation and the working together of collaboration, but fundamentally disrupt their explanatory power. Over this chapter and the next, I hope to show that collaborative work is not spontaneous, not based on individual merit, not without durable hierarchies, and especially not able to resolve conflicting points of view through novel policies. The first event I consider is a failed attempt to create a new article entry, focusing my attention on the debate about whether the article should stay or go. The second instance is a more predictable controversy over the inclusion of iconic images in the (already existing) article on Muhammad.

Article for Deletion: Wikipedia Art

> Yes, anyone can edit. No guarantee your edit will stick, though. All edits can also be reversed and deleted. Goes both ways, you see. So if you want to say Wikipedia is your temporary canvas, until someone notices what you did, then sure, it's your canvas.
> USER: EQUACZION

Wikipedia Art was a short-lived, highly controversial addition to Wikipedia. It was by no means a typical article, conceived rather as a work of concept art in the guise of an encyclopedic entry. The article was created on February 14, 2009, by the artists Scott Kildall and Nathaniel Stern, who describe their piece as an art intervention with "a nod to the traditions of concept- and network-based art" (Kildall and Stern 2011, 165).[1] The first few lines of the entry as it initially appeared on Wikipedia read:

> *Wikipedia Art* is a conceptual artwork composed on Wikipedia, and is thus art that anyone can edit. It manifests as a standard page on Wikipedia—entitled *Wikipedia Art*. Like all Wikipedia entries, anyone can alter this page as long as their alterations meet Wikipedia's standards of quality and verifiability. As a consequence of such collaborative and consensus-driven edits to the page, *Wikipedia Art* itself, changes over time.[2]

---

1. The artists' own account of Wikipedia Art can be found in *Critical Point of View: A Wikipedia Reader* (Kildall and Stern 2011).
2. *Wikipedia*, s.v. S. Kildall and N. Stern, "Wikipedia Art" (2009), accessed June 22, 2011, http://wikipediaart.org/wiki/index.php?title=Wikipedia_Artandoldid=211.

Wikipedia Art is not an immediately striking intervention; nor are its conceptual dimensions clear from the outset. At first impression, it also seems that ambiguities are only further augmented by the fact that the Wikipedia page, and therefore the art itself, could be edited by contributors other than the artists—ambiguity through participation is factored in to the piece itself. Wikipedia Art sits in the tradition of works such as John Cage's *4'33"*, where Cage famously performs nothing (split into three movements) for the duration of the work. One of the major themes of *4'33"* was the handing over of sound elements to the singularity of each performance. The emphasis was not on Cage's silence, but on the unpredictable sounds—a murmur, cough, creak, rustle, and so forth—that were heard during each performance. *4'33"* involved Cage setting up the structural conditions from which singular expressions of sound from any variety of sources became constitutive of the work. While not considered directly, Cage's piece is a very literal manifestation of what Umberto Eco famously described as "the open work" (1989). Like Cage's work, Wikipedia Art involves a stepping-back of the artist, embracing indeterminacy, and an exploration of structure vs. content. But also like *4'33"*, there is much more to Wikipedia Art than these factors and it would be a mistake to assume that the work doesn't speak in ways divorced from the particularities of its content. According to Kildall and Stern (2011, 167), Wikipedia Art is an intervention into the "power that Wikipedia holds, and the citation mechanism at the center of it all."

As is well known, the threshold for inclusion of an article on Wikipedia is not truth, but that an article or individual statement can be verified from a reliable outside source (see the policy entry titled "Wikipedia:Verifiability"). Kildall and Stern were of the opinion that while Wikipedia is merely supposed to document existing forms of knowledge, instead, "it often constitutes how we know the thing itself"; Wikipedia switches between "*record and source*" (170). Their idea was to expose the possibility of citation feedback circles, where an initial statement on Wikipedia could act as a *source* for a statement made outside Wikipedia by someone "reliable," which could then be used to verify the initial statement on Wikipedia.[3] To demonstrate this possibility, the creation of the Wikipedia Art article coincided with that of

---

3. Technically, the original statement on Wikipedia should probably never make it into the encyclopedia to begin with and could possibly be deleted as a case of "Original Research" or as having no sources to back the statement up (as the sources come after the article). However, it is precisely the messiness of the everyday workings of Wikipedia, where things are often overlooked, that the artists wish to highlight. It is also possible that the original statement could remain, and simply be marked as "needing citations."

several external publications, ones that satisfied the "reliability" criterion and could accordingly be used to verify Wikipedia Art into existence.[4] The artists frame Wikipedia Art as a demonstration of the performativity of language, where a statement can circulate and have real effects regardless of its truth value—though the artists also note that on Wikipedia such statements are often taken as true regardless of Wikipedia's explicit advice to the contrary.

Wikipedia Art no longer exists on Wikipedia. At present, the piece is archived at wikipediaart.org—a site that documents the unfolding of the "intervention" and includes related media coverage, links to discussions about the work, details of some of the participants, and documentation of the legal threats made by the Wikimedia Foundation for possible breach of trademark regarding the name "Wikipedia Art" by the domain wikipediaart.org.[5] There are, however, several traces of the entry that still haunt the site. At the former address (URL) of the original Wikipedia Art webpage, readers are presented with a short message about the article's (non)existence:

> This page has been deleted. The deletion and move log for the page are provided below for reference.
> • 06:30, 15 February 2009 Werdna (talk | contribs) deleted "Wikipedia Art" (*A7: No indication that the article may meet guidelines for inclusion*).[6]

The details reveal that the entry lasted a mere day before being deleted by Wikipedia administrator "Werdna." Besides these details and the links to information about Werdna are details (in brackets and with links) about why

---

4. It is worth noting that the artists document two instances—the article for Joy Division singer Ian Curtis, and the article for Digital Dark Age—where this possibility actually occurred. Wikipedia commentator Dror Kamir has written about similar though not identical processes of what might be called the co-constitution of events with Wikipedia articles relating to the Egyptian dimension of the Arab Spring. Kamir notes how the article "2011 Egyptian Revolution" was uploaded only a couple of hours after the actual protests began. Due to the fact that the upload process lasted only one minute, Kamir also concludes that the article was prepared in advance. It is well known that "social media" played a role in communicating and organizing the events and Kamir suggests that the Wikipedia article was part of this social media strategy. D. Kamir, "Parallel 'Online' and 'Real World' Egyptian Revolutions, or Wikipedia's Tahrir Square," July 1, 2011, http://anduraru.wordpress.com/2011/06/25/parallel-online-and-real-world-egyptian-revolutions-or-wikipedias-tahrir-square/.

5. Acting on behalf of the foundation, attorney Douglas Isenberg suggested that Kildall (who owns the Wikipediaart.org domain name) might be in breach of several trademark related acts, although the language used is unsurprisingly murky. Isenberg requested Kildall transfer the domain wikipediaart.org to the Wikimedia Foundation and to cease using the name Wikipedia Art to describe their activities.

6. *Wikipedia*, s.v. "Wikipedia Art," accessed January 30, 2012, http://en.wikipedia.org/wiki/Wikipedia_Art.

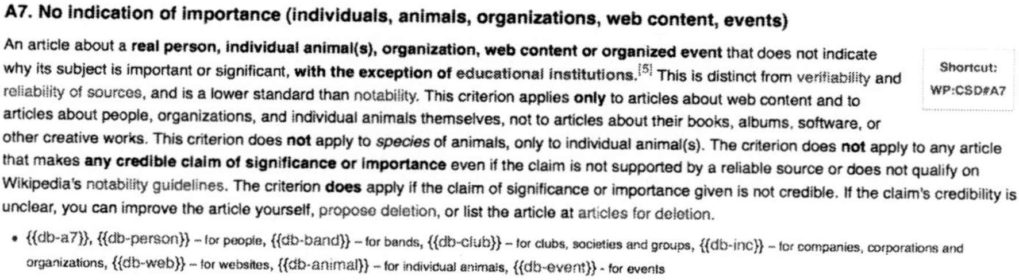

> **A7. No indication of importance (individuals, animals, organizations, web content, events)**
>
> An article about a **real person, individual animal(s), organization, web content or organized event** that does not indicate why its subject is important or significant, **with the exception of** educational institutions.[5] This is distinct from verifiability and reliability of sources, and is a lower standard than notability. This criterion applies **only** to articles about web content and to articles about people, organizations, and individual animals themselves, not to articles about their books, albums, software, or other creative works. This criterion does **not** apply to *species* of animals, only to individual animal(s). The criterion does **not** apply to any article that makes **any credible claim of significance or importance** even if the claim is not supported by a reliable source or does not qualify on Wikipedia's notability guidelines. The criterion **does** apply if the claim of significance or importance given is not credible. If the claim's credibility is unclear, you can improve the article yourself, propose deletion, or list the article at articles for deletion.
>
> Shortcut:
> WP:CSD#A7
>
> • {{db-a7}}, {{db-person}} – for people, {{db-band}} – for bands, {{db-club}} – for clubs, societies and groups, {{db-inc}} – for companies, corporations and organizations, {{db-web}} – for websites, {{db-animal}} – for individual animals, {{db-event}} - for events

FIGURE 3. Screenshot of "Criterion A7 for Speedy Deletion." *Source*: http://en.wikipedia.org/w/index.php?title=Wikipedia:Criteria_for_speedy_deletion&oldid=578472021 (accessed October 23, 2013).

the page was deleted. Following the link to "A7" takes readers to the policy page "Wikipedia:Criteria for speedy deletion." The page provides a list of criteria for when it is acceptable for Wikipedia administrators to "bypass deletion discussion and immediately delete Wikipedia pages or Media." The rationale for the existence of this administrator privilege is to "reduce the time spent on deletion discussions for pages or media with no practical chance of surviving discussion." The list of criteria include things like "patent nonsense," "pure vandalism and blatant hoaxes," "creations by banned or blocked users," "no context," "no content," and in the case of Wikipedia Art, "no indication of importance" (see fig. 3).[7]

However, before Werdna had swooped in and "speedily deleted" Wikipedia Art, thus classifying it as having no importance and "no practical chance of surviving discussion," a discussion about its merits had already begun. When there is a significant debate underway about the validity of an article, it is usually nominated as an "article for deletion." The nomination activates a series of procedures and rules for conducting and settling debates about deletion, which are outlined in the "Wikipedia:Articles for deletion" page. Any previous debate about the article's validity (usually from the "discussion" section of an entry) is copied over to a newly designated page where the rest of the debate plays out. "Wikipedia:Articles for deletion" further notes that "articles listed are normally discussed for at least seven days, after which the deletion process proceeds based on community consensus."[8] Wikipedia Art therefore followed a somewhat unusual trajectory, seemingly proving worthy of discussion and speedy deletion at the same time.

7. *Wikipedia*, s.v. "Wikipedia:Critieria for speedy deletion," accessed August 5, 2011, http://en.wikipedia.org/wiki/Speedy_Delete.
8. *Wikipedia*, s.v. "Wikipedia:Articles for deletion," August 5, 2011, http://en.wikipedia.org/wiki/Wikipedia:AfD.

While Kildall and Stern use Wikipedia Art to explore questions of performativity (something that I too explore in the following chapter, but in quite different ways), I am less interested in the flaws of Wikipedia's citation mechanism or how it functions as both source and document of authoritative knowledge. Instead, I want to consider Wikipedia Art in relation to questions about what constitutes collaborative knowledge on Wikipedia and the processes for filtering what gets included and what remains outside. I should also note from the beginning that I am not at all interested in whether or not Wikipedia Art should have remained or whether it was right to delete it. It is not the question of morals, of what is right and wrong, but the procedures of exclusion that in turn constitute an identifiable inside (an encyclopedia) that is worth pursuing.

At the bottom of the criterion "A7" there is a link to "Wikipedia:Articles for deletion" and the suggestion that when a claim has been made that an article is considered important, but the "credibility" of such claims is not clear, the article should be marked as an "article for deletion" and follow the set of related procedures. As noted above, articles for deletion (or AfD) "is where Wikipedians discuss whether an article should be deleted" and the aim is to arrive at "community consensus." There are links to a variety of related pages, such as "Wikipedia:Deletion process," "Wikipedia:Guide to deletion," and "Wikipedia:Deletion policy," which all cover different aspects of the deletion process, such as what constitutes grounds for deletion, the different kinds of procedures available, and advice about choosing the appropriate procedure.

As well as these general procedures, the AfD page provides more specific and technical instruction about listing a page for deletion and creating the separate discussion page. These technical details are followed by specific etiquette advice. The "AfD Wikietiquette" section rehashes existing etiquette guidelines and provides further deletion-specific advice. The Wikietiquette guideline regarding voting and consensus is worth highlighting: "Remember that while AfD may look like a voting process, it does not operate like one. Justification and evidence for a response carries far more weight than the response itself. Thus, you should not attempt to structure the AfD process like a vote."[9] It is a reminder that "consensus" on Wikipedia is not the result of typical democratic procedures. While the result of consensus might be general agreement between people, consensus is not primarily located in the agreement itself. Rather, consensus seems to exist beyond the agreement, outside it, in the realm of "justification and evidence." Obviously these things

9. Ibid.

are brought into the discussion, but they must, by definition, reside elsewhere. Although it may seem that this is a minor distinction, it isn't. It suggests that while Wikipedia tries to jettison questions of truth in some places, they pop up in others. While policies are designed to sidestep the problem of truth, these very policies must themselves be based on a truth—a truth of what is notable, verifiable, neutral, and what is none of these things and hence deleted. The capacity for participants to act in ways that can shape a dispute is intimately connected with these truths; it is subject to and derived from them. (I consider this question of truth in much more detail in the following chapter.)

Immediately after the etiquette pointers are instructions about "How to discuss an AfD." The opening passages read:

> AfDs are a place for rational discussion of whether an article is able to meet Wikipedia's article guidelines and policies. Reasonable editors will often disagree, but valid arguments will be given more weight than unsupported statements. When an editor offers arguments or evidence that do not explain how the article meets/violates policy, they may only need a reminder to engage in constructive, on-topic discussion.[10]

Following the same rhetoric as above, gestures are made to "rational discussion" and it is noted that although "reasonable editors" might disagree, the disagreement itself is thought to be resolvable through things usually associated with reason, such as evidence and valid arguments. Thus, although reasonable people might disagree, the hope is that the disagreement itself is not reasonable, that reasonable people can make unreasonable assertions, and that reason itself will emerge through its own contestation. At the bottom of the AfD page is a search bar that provides access to the archive of all previous AfD discussions. It is here that the most important trace of Wikipedia Art resides: the record of the debate itself. The deletion debate was quite short, both in length and time (roughly 7,500 words over one day), but it nonetheless makes visible how the instructions for deletion discussion noted above actually played out, and it shows the failed attempts by supporters of Wikipedia Art to "verify it into existence."

As noted on the AfD "how to" section, discussants generally begin their contribution with a pronouncement of what they think the fate of the article should be, or by classifying their contribution in an immediately graspable manner. The most common classifications are "keep" or "delete," but in this particular debate others include "comment," "proposal," "recap," and "move

10. Ibid.

to project space." The deletion discussion is opened by user DanielRigal, the same user who marked it as an AfD. This user writes:

> This is an attempt to use Wikipedia as an "art platform." It is not encyclopaedic. It can never be encyclopaedic by its very nature. It can't be referenced to anything other than itself because it is an original work based on Wikipedia. These guys need to get themselves their own Wiki and host this there. It also seems to be part of a walled garden of suspicious articles about the artists themselves (Scott Kildall, Nathaniel Stern, and Brian Sherwin). It seems that they have accounts and edit these themselves. They may, or may not, be significantly notable outside of their own circle and may, or may not, have inflated their importance in their articles. I think it needs looking at. DanielRigal (talk) 20:54, 14 February 2009 (UTC)[11]

There are two main arguments put forward and several issues raised in DanielRigal's initial post. The first and most obvious criticism is that it is not an encyclopedic contribution. While DanielRigal does not explicitly refer to any policies, guidelines, or principles, this first argument is supported by the first of Wikipedia's Five Pillars: "the fundamental principles by which Wikipedia operates."[12] The first principle states that

> Wikipedia is an online encyclopedia. It incorporates elements of general and specialized encyclopedias, almanacs, and gazetteers. Wikipedia is not a soapbox, an advertising platform, a vanity press, an experiment in anarchy or democracy, an indiscriminate collection of information, or a web directory. It is not a dictionary, newspaper, or a collection of source documents; that kind of content should be contributed instead to the Wikimedia sister projects.[13]

DanielRigal's second argument leads directly from the first and further serves to define what constitutes something as "encyclopedic." Wikipedia Art cannot be encyclopedic, the argument goes, because it only exists on Wikipedia and therefore "can't be referenced to anything other than itself." It is an argument about "verifiability" and serves to define "encyclopedic" in such terms. As noted above, it is also the connection between or rather definition of *encyclopedic* as that which is verifiable that the artists wish to interrogate. The post finishes by flagging concerns about self-editing (which relates to

---

11. Unless otherwise indicated, all cited material from the AfD discussion is from, *Wikipedia*, s.v. "Wikipedia:Articles for deletion/Wikipedia Art" (2009), accessed August 5, 2011, http://en.wikipedia.org/wiki/Wikipedia:Articles_for_deletion/Wikipedia_Art.

12. *Wikipedia*, s.v. "Wikipedia:Five pillars," accessed November 7, 2011, http://en.wikipedia.org/wiki/Wikipedia:Five_pillars.

13. Ibid.

the behavioral guideline about "conflict of interest") and by questioning the notability of the artists themselves (see "Wikipedia:Notability").

The first to respond to DanielRigal is a user called Artintegrated, who begins by noting, "Whether these people do simple edits on their own pages in no way validates what they have said here. If something is true then it should stay in the article regardless." It is targeted loosely at the concerns that DanielRigal finished on. Following this, Artintegrated writes, "Did you know this article is already referenced at The Whole 9 . . . just today. I feel that your idea that it can only reference itself is unfounded at this point." This is an attempt to overcome the verification dilemma, and thus cuts to the heart of the Wikipedia Art experiment. DanielRigal immediately recognizes the issue and responds accordingly:

> you can't have a circular chain of references. You can't reference Wikipedia from a non-RS [reliable source] blog that itself references Wikipedia. By that logic, any information replicated on two different websites and referencing each other would be gospel truth. Referencing does not work like that.

DanielRigal also notes that users can't write their own articles because they "lack objectivity." Two more users add comments: One responds to the objectivity question, "there is no such thing as objectivity on Wikipedia. That is the whole point—it is inherently subjective," and the other suggests giving the article "time to improve." To this DanielRigal responds,

> Please read the article carefully and see that it can't possibly improve to become a valid Wikipedia article. It is an article about itself. It is intrinsically unencyclopaedic. I don't think it was necessarily created in bad faith but it is an abuse of Wikipedia to seek to use it as an art platform and it undermines Wikipedia as an encyclopaedia.

The early part of the debate therefore follows two lines, one on what is considered encyclopedic in relation to verifiability and the other on whether or not it is acceptable for editors to write material about themselves. And while DanielRigal is initially outnumbered three to one, new discussants soon come to his aid. RHaworth categorizes their post as Delete and writes, "Only fractionally better than any MADEUP topic. Created very recently. Also a totally confused concept—a collaborative art project—fine. But trying to do it on one Wikipedia page—you must be joking mate! We also have an avoid self-reference rule."[14] Contributors JohnCD and LtPowers also suggests

---

14. "MADEUP topic" is a reference to the content guideline "Wikipedia:Wikipedia is not for things made up one day" (http://en.wikipedia.org/wiki/Wikipedia:MADEUP) and that

deletion: "an interesting concept, but not suitable *here*: this is an encyclopaedia" and "Out of scope as a project, completely lacking in evident notability as a concept."

At this point DanielRigal discovers the artists' own wiki, which mirrors the page on Wikipedia. It forces him to revise his initial argument:

> OK. Now I am really confused. They have a Wiki of their own at: wikipediaart. org, which has the same content as the Wikipedia article we are discussing here. I am not sure how the two are meant to relate to each other but it may be that they are confused as to the difference between a Wiki and Wikipedia. I am not sure which site they are proposing to be the actual art work. If it is the Wikipedia article then all I have said above is correct. If it is their own Wiki then the circularity is broken and the article is not *intrinsically* unencyclopaedic. In that case I would like to add the following alternative reasons to delete the article: Lack of notability and lack of RS references.

It seems now that it isn't the very possibility of the article that is objectionable, but rather that it isn't notable enough and is still not verified by reliable sources. A discussion about the location of the art project and how that bears on the encyclopedia entry also follows. Freshacconci enters the debate by affirming DanielRigal's initial position, but then adds another layer of complexity:

> This could never be properly sourced, as it could only exist here first before it could ever be written about in order for it to be notable enough to be mentioned here. Yes, an interesting paradox, but that's not our problem. We can only go by Wikipedia policies and guidelines, and it's pretty clear that this needs to be deleted. But here's an idea: the fact that this was attempted and subsequently deleted could possibly generate enough third-party coverage to make the initial project notable enough to be included (at least as part of the artists' articles). But until then, it cannot stay. It's not encyclopedic as an entirely self-referential article.

And thus, the possibility of performing itself into existence on Wikipedia lives on.

By the middle of the debate there is still nothing close to consensus, at least as defined in the traditional sense of "agreement." New arguments continue

---

stipulates "Wikipedia is not for things that you or your friends made up. If you have invented something novel in school, your garage, or the pub, and it has not yet been featured in reliable sources, please do not write about it in Wikipedia." The self-reference rule RHaworth refers to is part of the Manual of Style guidelines. It advises contributors not to refer specifically to Wikipedia when writing articles.

to be introduced, while some points are labored many times over. Statements in favor of deletion come to include:

> "This does not make any sense: it is an article about itself. I think the article is a breaching experiment."
> "This does not fit Wikipedia."
> "'Wikipedia Art' fails WP:N and WP:V."
> "I see no reason to make an exception for its failure to meet basic requirements for Wikipedia articles. In the absence of any reasons given for overriding Wikipedia basic policy, I see no reason not to delete 'Wikipedia Art.'"
> "Previous discussions about sourcing are besides the point, because this is an art project, and art projects are not allowed in article space."
> "an article is an attempt to objectively capture the facts about a subject and . . . art is a subjective attempt to say something original about something. Given that Wikipedia is for objectivity and against original research it really is an incredibly inappropriate place to seek to make art."
> "We ask for reliable sources and you give us blogs. We complain of original research and you seek to remedy it by soliciting more original research. I would have expected better."
> "Speedy Delete—G1, G2, G3, or G11—Take your pick. How about simply not notable, vandalism, hoax, etc? Whether it can be considered art or not is irrelevant. Wikipedia ain't your canvas."
> "This 'article' seems designed to violate as many of our basic policies as possible. Linking every word? Signatures in article space? Ridiculous amounts of self-referencing? An article that is about nothing but itself? It is absurd."
> "WP:OR, WP:SOAPBOX, not notable, no reliable sources except one blog, trying to use wikipedia for something other than writing an encyclopedia. . . . why are we even having this discussion?"

Finally, there is a suggestion that Wikipedia Art is "most likely infringing on the Wikimedia Foundation's copyright on the name *Wikipedia*."

While there are, at least in the middle of the debate, equal voices in favor of keeping Wikipedia Art, the mode of argumentation is notably different. The excerpts show how "Deleters" regularly refer to policies and guidelines and how they tend to be highly dismissive of the article/artwork. For their part, the "Keepers" rarely refer to established policies and guidelines to support their claims. Their argumentative mode is far more deconstructive and explorative, often challenging or attempting to redefine existing rules. For example, in response to the charge that Wikipedia "is not a web host for collaborative art projects" an unsigned user questions, "What exactly distinguishes a collaborative art project from a collaborative article?" In a similar vein, Shmeck provides a lengthier contribution:

> The Wikipedia Art page is something that explains art, explores art, and is art all at the same time. Deleting this page would be a statement that the exegesis of conceptual art and/or new media art has no place in Wikipedia, except on the tired, lifeless, and opaque conceptual art and new media art pages. Why shouldn't a tiny corner of Wikipedia-brand collective epistemology be preserved for an instructive, self-referential, and ever-changing living example of what an art object can be in the 21st Century? Should this page be judged invalid only because it refers to itself? The Wikipedia Art page is a self-aware example of Wikipedia's mission of collective epistemology. It enacts and exposes Wikipedia's own strengths, weaknesses, potential, and limits as a system of understanding and as a contemplative object of beauty. The page is also a self-aware example of the strengths, weaknesses, potential, and limits of new media art as an object of contemplation. New media art has demonstrated that the boundaries between art and every other discipline from epistemology to microbiology have disintegrated (see interdisciplinarity) in the 21st Century. This page shows how a Wikipedia page can go beyond simply existing as a Wikipedia page, while retaining its basic utilitarian Wikipedia function. Those who care most about Wikipedia's mission would probably agree that Wikipedia already is a collaborative art form. If you feel that Wikipedia is a beautiful thing, then at some level (whether or not you admit it) you consider Wikipedia an art form, with its own codes and conventions. This artwork can only exist as a Wikipedia page that refers to itself. Therefore, deleting would not only send the message "this is not Wikipedia"; it would also be saying "this is not art."

The contribution tries to bridge the gap between art and encyclopedic knowledge that underpins many of the Deleters' arguments. This is done by a two-way interrogation: First, Shmeck points to the educational value of art: While Wikipedia Art might not be strictly encyclopedic, it is nonetheless educational, which is the ultimate function of an encyclopedia. The strategy is then reversed and Wikipedia is described as "a beautiful thing." The specific argument is that to deny the existence of Wikipedia Art is to deny the beauty and hence the aesthetic value of Wikipedia as a whole.

These kinds of argumentative strategies and attempts to redefine the terms of debate lead DanielRigal to make the following reflective comment:

> Recap: I think we have an unusual situation here in two ways. First up there are a lot of people here who do not normally "do" AfDs. Secondly, there is a real, and I believe honest, failure of those who want to keep the article to understand the fundamental nature of the problem, or of Wikipedia itself. I don't want to be patronising but lets quickly recap Wikipedia 101: The five pillars of Wikipedia explains what Wikipedia is, isn't and also how it is run. Almost everything of importance is linked from there but I would specifically

like to mention notability, verifiability, reliable sources, no original research and, last but not least, do not disrupt Wikipedia to illustrate a point.

Immediately following this comment are two attempts by Keepers to mobilize, rather than critique or redefine, existing rules. Both Patlichty and Shane Mecklenburger mount arguments for "notability" and "verifiability" and the latter addresses issues of "reliable sources," "no original research," and the "do not disrupt Wikipedia to illustrate a point" behavioral guideline. Once again, though, the Keepers refer to these rules in highly strategic ways or in a manner otherwise deemed unacceptable by the Deleters. Patlichty, for example, uses his own status as a "New Media Art professor and curator" as part of his argument about notability, which is quickly pointed out and dismissed by DanielRigal, who soon after proposes to close the entire discussion in favor of delete.

Although one contributor notes closing the discussion "within the first couple of hours" is not standard practice and suggests "this is way too soon in the process for this to happen unless the person who put it up for deletion is afraid that those of us who support the article will ultimately see the page remain," the final part of the discussion is a flurry of suggested deletions.[15] Finally, the administrator called Werdna answers DanielRigal's request. Werdna ends the discussion with this statement: "Speedily deleted. No indication that the content may meet our criteria for inclusion." At the same time Werdna deletes the actual Wikipedia Art article, leaving a very similar statement about inclusion (noted above) and a link to the A7 criterion for speedy deletion. Thus ended the life of the entry on Wikipedia Art.

Images of Knowledge: Muhammad

There are many controversial articles on Wikipedia where conflict is manifested not in relation to an article's very existence, as in the case of Wikipedia Art, but in regard to one aspect of its content. The site itself provides a long list of "controversial issues," which are sorted into fifteen categories. The opening lines of the page "Wikipedia:List of controversial issues" defines a controversial issue as "one where its related articles are constantly being re-edited in a circular manner, or is otherwise the focus of edit warring" and further states:

---

15. The most amusing is Huntster's contribution: "Delete as non-notable, self-referential mess. Tried by others, and deleted. Kill kill kill."

> This page is conceived as a location for articles that regularly become biased and need to be fixed, or articles that were once the subject of an NPOV dispute and are likely to suffer future disputes. Other articles not yet classified as "controversial" have some edit conflict issues. The divisive nature of disputed subjects have triggered arguments, since opinions on a given issue differ as they are debated. These subjects are responsible for a great deal of tension among Wikipedia editors, reflecting the debates of society as a whole. Perspectives on these subjects are particularly subject to time, place, and culture of the editor.[16]

A controversial issue is therefore one that creates a kind of impasse, which is framed in terms of bias and where problematic articles are unable to stabilize a neutral point of view. Many of the articles listed are highly predictable, including such topics as 2003 invasion of Iraq, abortion, American Jews, capital punishment, neo-liberalism, genocide, stem cell research, Scientology, age of consent, polygamy, global warming, PETA and animal rights groups, deforestation, and tax. But these exist alongside less obvious articles, such as Kid Rock, Yuppies, Netiquette, and 69ing.

Likewise, the "humorous" Wikipedia page "Wikipedia:Lamest edit wars" provides a long list of less typical examples that have, for one reason or another, generated lengthy debates and edit wars.[17] Like the serious list of controversial topics, the page also categorizes its articles. While some of these categories overlap with those found on "Wikipedia:List of controversial issues," such as politics and religion, the page also includes categories that are much more specifically about creating encyclopedic knowledge, such as names, dates, numbers and statistics, spelling, punctuation, and wording. One specific example that scholars of new media might appreciate is whether or not to capitalize the "d" and the "b" in danah boyd.[18] Thus, the spectrum of controversy on Wikipedia is wide, reflecting long-standing issues outside the encyclopedia and generating more specific and mundane ones in the process of translating seemingly noncontroversial topics into the genre (or what I will later call the "frame") of the encyclopedia.

---

16. *Wikipedia*, s.v. "Wikipedia:List of controversial issues," accessed August 5, 2011, http://en.wikipedia.org/wiki/List_of_controversial_articles.

17. A visualization of a selection of these articles, detailing the number of edits, category of article, and an approximation of the duration of each edit war is available at www.informationisbeautiful.net/visualizations/wikipedia-lamest-edit-wars/.

18. *Wikipedia*, s.v. "danah boyd," accessed July 8, 2011, http://en.wikipedia.org/wiki/Danah_boyd.

The article entry for Muhammad is not featured on Wikipedia's list of controversial topics, or within the inglorious history of lame edit wars, but its history of edits makes it a worthy candidate for any such lists. It also shares much in common with many of the topics listed as controversial, especially in terms of reflecting controversies outside the encyclopedia. For example, edit activity for Muhammad peaked in the aftermath of the global controversy sparked by the publication of cartoon depictions of Muhammad by the Danish newspaper *Jyllands-Posten*.[19] The entry was created on November 8, 2001. As of July 2011, it has been edited just under 17,000 times, by 4,033 users, and is roughly 16,000 words long including references.[20] The top 10 percent of active users have contributed 10,951 edits, which accounts for a little over 65 percent of the total. On average, the page is edited 145 times every month. These introductory statistics make clear that the entry for Muhammad is a large and heavily edited entry. It fits within the wider paradigm of participation; is an obvious work of collaboration, especially during the period of controversy when editing activity was way up; and is demonstrative of some of the other concepts introduced at the beginning of this chapter (e.g., stigmergy, produsage, modularity).

The page itself presents like any other Wikipedia entry, with a short introductory paragraph, followed by a table of contents that also serve as links to specific parts of the page. Running alongside the table of contents, to the right, is a green box indicating that the article is "part of the series: Islam." Those interested in Islam can expand and click on links within this box to further delve into the topic. The written component of the article does its best to adopt the dry, formal, and descriptive style that has come to characterize the genre of the encyclopedia. There are sections on the life of Muhammad; on his legacy, wives, and children; his dealings with slaves; the main sources of knowledge about him; and a small section on criticism. There are timelines, links to related pages, a list of references and notes, further reading,

---

19. The cartoon depictions were first published on September 30, 2005, under the headline "Muhammeds ansigt" (The face of Muhammad). The cartoons were published as a response to a perceived culture of self-censorship that had emerged in Denmark on the topic of Islamic criticism. News of the depictions spread quickly throughout the globe and led to protests in many Muslim nations. The Danish embassy in Pakistan was bombed, and in Syria, Lebanon, and Iran the embassies were set on fire. In Nigeria the BBC also reported that "Danish cartoon protests in the north led to sectarian clashes which have seen dozens of deaths in four cities" (BBC News 2006).

20. These statistics and the ones that immediately follow were generated using the "Page History Statistics" tool, available at toolserver.org/~soxred93/articleinfo/.

and eighteen images in total (including maps), six of which are depictions of Muhammad. And while the page is constantly changing in small ways and occasionally larger ones, it nonetheless presents as a relatively fixed entry. The entry has achieved a level of stability that allows readers to approach its content with a certain amount of predictability; the material is identifiably encyclopedic and thus bears no mark of controversy or conflict. To be clear, it is not that encyclopedias do not cover controversial material but rather that they present such material from a position that has itself been rid of conflict and controversy. In the past there have been a variety of strategies for unifying a field of knowledge, for laying claim to the incontestable truth of something, and we have already seen (and will see in more detail later) that Wikipedia's primary tools in this regard include the NPOV policy, together with no original research (NOR) and verifiability.

One aspect of Wikipedia that sets it apart from its historical precedents, however, is that the article page (the one that contains the typical encyclopedic content) is only one component of the entry. The article page, also described as the "project page," sits alongside a discussion page (or pages), an edit page (from which edits to the project page can be made), and a history page (where every prior version of the page can be accessed). The article page is very much only the tip of the iceberg of any particular entry and the rest of the entry presents an often radically different side of Wikipedia, one where heated debates and edit wars are not uncommon. The history logs make visible transformations in content, forms of vandalism, and things like "revert wars" where two or more parties keep reverting each other's edits. But the richest content lies in the discussion pages. The discussion page "Talk:Muhammad" begins with a series of meta information boxes about the article and related discussion. The first is labeled "Important notice" and features a stop hand icon (fig. 4). It is a notification about the status of depictions of Muhammad. Below this "Important notice" box are six other boxes, some of which require expanding (clicking on a "show" button) to reveal the content. Together with the first box, several of these boxes act as a kind of official defense against anticipated future complaints about the images. A small box reminds editors about neutrality and flags the article as sensitive: "Please be neutral when editing this highly sensitive article. It discusses a topic about which people have diverse opinions." There is also a Frequently Asked Questions (FAQ) box, which includes carefully crafted responses to questions such as "Shouldn't the images be removed because they might offend Muslims?," "Aren't the images false?," and "Can censorship be employed on Wikipedia?" This box is very revealing about the entry as a whole. It immediately gives a sense of past debates, of the kind of perspectives or "points of view" that have

> **Important notice**: Prior discussion has determined that ***pictures of Muhammad are allowed and will not be removed from this article***. **Discussion of images should be posted to the subpage** Talk:Muhammad/Images. **Removal of pictures without discussion will be reverted.** If you find Muhammad images offensive, it is possible to configure your browser or use your personal Wikipedia settings not to display them, see Talk:Muhammad/FAQ.
>
> The **FAQ** below addresses some common points of argument, including the use of images and honorifics such as "peace be upon him". The FAQ represents prior consensus of editors here. If you are new to this article and have a question or suggestion for it, please read the **FAQ** first.

FIGURE 4. Screenshot of "Important Notice Box for Entry on Muhammad." *Source*: http://en.wikipedia.org/w/index.php?title=Talk:Muhammad&oldid=446702716 (accessed August 25, 2011).

clashed in these debates, as well as information about the editing status of the article (which is blocked to anonymous editors and accounts less than four days old).

Most of the responses to these FAQs are oriented around existing policies and guidelines. Figure 5, for example, shows the response to the first FAQ. It begins with links to Wikipedia's policy about censorship and the site's general content disclaimer, which warns that Wikipedia contains content that may be considered objectionable. The response notes that Wikipedia is not "bound by any religious prohibitions" and further references the NPOV policy. The passage on censorship situates Wikipedia as beyond the editorial control and hence "benefit of any particular group." The last section of the response tries to demonstrate this and therefore ward off accusations of an anti-Islamic perspective by providing similar non-Islamic examples of these policies at work. What these boxes and FAQ responses demonstrate is not that Wikipedia doesn't discriminate between specific groups and the forms of knowledge that help constitute them. Rather, and as I will elaborate later, Wikipedia is defined through the systematic exclusion of certain forms of knowledge—and it does not discriminate between these discriminations!

Contributing to talk pages on Wikipedia is generally quite straightforward. A person simply adds a new topic to the existing list and writes down his or her concerns or discussion points. To add to an existing topic, a contributor simply clicks on the "edit" button on the top-right of the topic heading and then adds comments, which appear slightly indented and below the previous comments. As of July 2011, there are nine current topics listed for discussion. In spite of all the instructions and reminders found in the boxes, three of the topics for discussion are "edit requests" to remove the images of Muhammad and thus not really topics for discussion at all. Edit requests

> **Q1: Shouldn't all the images of Muhammad be removed because they might offend Muslims?** [hide]
>
> **A1:**
>
> > *Further information: Wikipedia:What Wikipedia is not#Wikipedia is not censored and Wikipedia:Content disclaimer*
>
> There is a prohibition of depicting Muhammad in certain Muslim communities. This prohibition is not universal among Muslim communities. For a discussion, see Depictions of Muhammad and Aniconism in Islam.
>
> Wikipedia is not bound by any religious prohibitions. Wikipedia is an encyclopedia that strives to represent all topics from a neutral point of view, and therefore Wikipedia is not censored **for the benefit of any particular group.** So long as they are relevant to the article and do not violate any of Wikipedia's existing policies, nor the law of the U.S. state of Florida, where most of Wikipedia's servers are hosted, no content or images will be removed from Wikipedia because people find them objectionable or offensive. (See also: Wikipedia:Content disclaimer.)
>
> Wikipedia does not single out Islam in this. There is content that may be equally offensive to other religious people, such as the 1868 photograph shown at Baháʼuʼlláh (offensive to adherents of the Baháʼí Faith), or the account of Scientology's "secret doctrine" at Xenu (offensive to adherents of Scientology), or the account at Timeline of human evolution (offensive to adherents of Young Earth creationism). Submitting to all these various sensitivities would make writing a neutral encyclopedia impossible.

FIGURE 5. Screenshot of "Response to FAQ 1 for Entry on Muhammad." *Source*: http://en.wikipedia.org/w/index.php?title=Talk:Muhammad&oldid=446702716 (accessed August 25, 2011).

are made on Wikipedia discussion pages "where the editor cannot or should not make the proposed edit themselves."[21] In the case of the entry on Muhammad, edit requests are made because these particular users, being either anonymous or very new, cannot edit the page due to its semi-protected status. "Edit requests" manifest a kind of stubborn protest in the face of powerlessness; they are an attempt to resurrect an issue that is otherwise trying hard to remain dead and buried. In order to see how the issue over images was initially settled—a task that actually proves very difficult—one has to consult the discussion archives.

The archives for Talk:Muhammad are extensive. They are broken into three categories: general, nontitled archives (25 separate pages); dedicated image archives (18 pages); and mediation archives, which document attempts to mediate a specific issue (9 pages). All of the mediation archives deal with the images of Muhammad. The general archive contains innumerable topics on all aspects of Muhammad as well as general discussion about how to interpret policies in light of the material at hand. The first archive, for example, which begins in February 2005, includes discussions about the deletion of statements about Muhammad being poisoned before death and then moves to consider how his death should be depicted more generally; concerns of "serious bias" in statements about Ali; reverting "the prophet passes" to "The Death of Muhammad" and the related issue of religious neutrality; the general arrangement of the text; whether or not it is appropriate to have a

---

21. *Wikipedia*, s.v. "Wikipedia:Edit requests," accessed August 5, 2011, http://en.wikipedia.org/wiki/Wikipedia:Edit_requests.

dedicated criticisms of Muhammad section in the article;[22] whether or not the edit war about the "Muhammad as warrior" section stems from Islamophobia; how best to describe young males who were beheaded after a battle; whether or not Muhammad married Aisha when she was six years old and had intercourse with her at the age of nine; why the addition of PBUH after any mention of Muhammad is not suitable in Wikipedia;[23] whether describing him as poor and illiterate is suitable; and there are many concerns raised about the lack of NPOV, of bias, and the overall poor quality of the article.[24]

Thus, while the majority of discussion about the Muhammad entry concerns the images, almost every knowledge statement that has made it into the entry has undergone some kind of scrutiny. Questions of order, orientation, tone, style, genre, length, and perspective; of what is to constitute Muhammad as a knowledge artifact within a specific encyclopedic paradigm, and in turn what kinds of statements characterize the encyclopedia more generally: this is what the talk pages make visible. Any relatively established article entry, stripped of visible signs of conflict and controversy, must therefore be seen as the *outcome* of discussions and battles, of the competition of statements that takes place on discussion pages.

It is not possible to cover every debate that has taken place about depictions of Muhammad, nor is it necessary as there is much repetition—a fact that is itself significant, as we shall soon see. Almost every aspect of the images is debated in these archives: whether or not the images should be included at all; which ones to include; their position, number, accompanying caption and so on. As much as possible, I will focus only on the question of whether the images should be included at all. To give a sense of this debate, in appendix A I have extracted a selection of statements from the first ten general archive pages. These are categorized into positions for inclusion, against inclusion and attempts at some kind of compromise or novel solution. I have further categorized these around the main themes of the statements, such as censorship, relevance, respect, and accuracy. This selection is by no means comprehensive; nor are the subcategories of special import. Often contributors will invoke several arguments in the one post, or present one statement that is implicitly based on another. The hierarchy of these subcategories also

22. User Mustafaa, who opens this topic, thinks that this section is unusual, considering, "Not even L. Ron Hubbard gets one of those, let alone Jesus."
23. PBUH is an acronym for "Peace be upon him." Practicing Muslims often say this after speaking or hearing the name of a prophet of Islam.
24. *Wikipedia*, s.v. "Talk:Muhammad/Archive 1," accessed August 5, 2011, http://en.wikipedia.org/wiki/Talk:Muhammad/Archive_1.

changes between different contributors and sometimes the same policies and themes are used to make opposing arguments. Each position is often accompanied by detailed counterarguments, and a game of picking apart arguments commences. What appendix A shows is how polarized the debate about the images is. It gives a sense of how arguments are put forward, how similar statements are invoked repeatedly, and how contributors relate to and position one another, as well as some common points of disagreement.

Whereas the FAQ box considered above is policy heavy, in the actual debate policy considerations are invoked less frequently (though they do remain visible and important in some sections). The selection also shows how in certain instances one finds statements in which the actual argument remains implicit. It is often enough simply to affirm a statement, such as 1.1: "Wikipedia is not censored." The themes of different statements vary from highly predictable and regular, such as censorship, religion, offensiveness, and encyclopedic value, to more novel ones such as an analogy about neighborly barbecue relations and vegetarianism (8.1) or the claim that having the images is offensive to Muhammad himself (as opposed to followers of the Islamic faith) (10.1).

Appendix B contains excerpts from the mediation archives. Mediation is a formal process in Wikipedia that involves bringing in a third-party Wikipedian who has no previous involvement with the entry.[25] The mediation under consideration began in November 2006, and was undertaken by a contributor named Aguerriero. After some initial clarification about exactly what is at stake (one particular image or all images), and someone flagging that the mediation had been "advertised" on a "partisan Muslim Guild,"[26] Aguerriero asks all participants in the debate to state their position as concisely as possible. Selections of these statements make up the first section of appendix B. These position statements cover much of the same ground as appendix A, except they all fall into categories of *for* or *against*, without the more ambiguous *solution* category I included in appendix A. After every contributor has put forward a position, Aguerriero tries to condense them all into two general positions and then asks contributors to indicate whether or not the summaries (which follow) accurately reflect the "major sides of the issue" (section 2 of appendix B):

25. See *Wikipedia*, s.v. "Wikipedia:Mediation Committee/Policy," http://en.wikipedia.org/wiki/Wikipedia:Mediation_Committee/Policy.
26. The contributor, Proabivouac, was concerned that a sudden influx of Islamic perspectives might deviously alter the outcome of the mediation. However, Aguerriero reminded Proabivouc that the desired outcome of mediation, consensus, is "not a 'vote,' so numbers of people will not matter."

SORTING COLLABORATION OUT

> Encyclopedic depictions of Muhammad should be included in the article. Removal on the basis of relevance or notability may be discussed on a per-image basis.
>
> Depictions of Muhammad should not be included in the article since they are offensive to many Muslims who read Wikipedia, and the depictions may be made available in a separate article (such as Depictions of Muhammad).

This is an attempt to identify the source of the disagreement, to filter out all nonessential differences and to subordinate all others to these terms. After the discussants weighed in on these two positions Aguerriero makes a second attempt to capture the essence of the debate:

> Encyclopedic depictions of Muhammad should be included in the article, and held to defined standards of notability and relevancy. Standards will be defined in this mediation.
>
> Depictions of Muhammad are not informative (and by extension, not encyclopedic) because the physical appearance of Muhammad is unknown, and the depictions are offensive to many Muslims. As such, the depictions should not appear in the article.

After further discussion and tweaking, Aguerriero suggests discussants try to establish the standards of inclusion for an image—as related to the first position—and provides suggestions (section 3 of appendix B). The suggestions spark a lengthy debate, which seems to lose focus and contributors revert to posts that resemble mini-essays. By page 6 of the archive it seems as though Aguerriro has left the mediation and there is an attempt to begin anew. Many contributors chime in, to little avail and there is an attempt to take a "poll" on each of the issues at hand to get a sense of where people stand. By archive 8 there is a suggestion to take the case to Wikipedia's Arbitration Committee (which doesn't usually handle "content disputes"), and one editor titles his or her contribution "This mediation is going nowhere." At the end of archive 8 there is a vote regarding whether or not the mediation itself was a success, which also renders no clear result. The mediation ends with a post titled "A Solution: The End?," which states the following: "Right now after User:Alecmconroy edit we have a compromise. Please vote for it at Compromise Found" (the words "Compromise Found" is a link). However, this "Compromise Found?"[27] post in the regular archive also includes a vote, with differing positions, and has clearly not reached "consensus."

---

27. The post is archived at http://en.wikipedia.org/wiki/Talk:Muhammad/images/Archive_1#Compromise_found.3F.

The archives show that the debate about depictions of Muhammad was long and messy. The model of consensus, based on reason instead of voting numbers, did not lead to general agreement and the mediation attempt was similarly unsuccessful (despite claims to the contrary). While discussion activity might have settled down somewhat, the *debate itself* was never conclusively settled, nor is it entirely dead as the ongoing requests for deletion make clear. In light of this, the boxes that sit atop the discussion archives, which signal to editors that the debate about images has been settled, that "prior discussion has determined that *pictures of Muhammad are allowed and will not be removed from this article*," are somewhat misleading. And the same is true for the neatly reasoned, policy-informed responses to the FAQs. Wikipedia Art and the debate about the inclusion of images of Muhammad make visible quite different dimensions of Wikipedia than those usually mobilized in discussions of participation and collaboration. For the remainder of this chapter, I use these instances as the raw material from which to explore the political dimensions of organizing contributions (production) in Wikipedia.

Frames

The debate about Wikipedia Art and the inclusion of images in Muhammad take place on two distinct levels of the encyclopedia and they involve two equally distinct though related political moments—moments that invoke what I call *a politics of the frame*. As a concept, the frame has a rich and diverse theoretical history, beginning perhaps with Alfred Schutz's phenomenological sociology (1970). My own consideration begins with a short essay by Gregory Bateson titled "A Theory of Play and Fantasy," which was later republished in *Steps to an Ecology of Mind* (Bateson 1972, 177–93). Bateson uses the concept of the frame to explore the relation between abstract metacommunication and "psychiatric theory." Among other things, Bateson was interested in those aspects of communication that also signal something more than the message or, rather, that provide signals *about signals*—about how a message is to be understood. In particular, Bateson considers the question of play and how it is that human and nonhuman animals can recognize a series of signals as such. He invokes two useful analogies that help mark an entry point into thinking about frames. The first is that of a diagram used in set theory, where items are organized into specific sets in relation to axioms or basic principles. The principles define which items are deemed meaningful and belong in the set and which are not and are thus relegated to the outside of the frame. In terms of play, the set would include all of the statements between two human or nonhuman animals that can be classified as such (as play) within a specified duration. Bateson describes such set theory diagrams therefore as "a topological approach to the logic of classification" (186). From the outset, then, a frame is a mode of referring *by ordering*. A frame always sorts things as either belonging or not belonging and this process is mediated by axioms or principles—indeed, the axioms are what define the frame; they are the conditions of its possibility.

The second analogy Bateson employs is the picture frame, which is not considered on its own but through a dialogue with the first analogy and in the process of identifying the "common functions" of framing in general.[28] In addition to "excluding" and "including" certain messages or "meaningful actions" (which the set theory analogy highlights), frames serve an interpretive or perceptive function and mark a qualitative distinction between what is included and what is left out:

> The frame around a picture, if we consider this frame as a message intended to order and organize the perception of the viewer, says, "Attend to what is within and do not attend to what is outside." Figure and ground, as these terms are used by gestalt psychologists, are not symmetrically related as the set and non-set of set theory. Perception of the ground must be positively inhibited and perception of the figure (in this case the picture) must be positively enhanced. (1972, 187)

Bateson continues:

> The picture frame tells the viewer that he is not to use the same sort of thinking in interpreting the picture that he might use in interpreting the wallpaper outside the frame. Or, in terms of the analogy from set theory, the messages enclosed within the imaginary line are defined as members of a class by virtue of their sharing common premises or mutual relevance. The frame itself thus becomes a part of the premise system. Either, as in the case of the play frame, the frame is involved in the evaluation of the messages which it contains, or the frame merely assists the mind in understanding the contained messages by reminding the thinker that these messages are mutually relevant and the messages outside the frame may be ignored. (187–88)

In addition to sorting and ordering, frames are lenses of differentiation; they mark qualitative (as well as quantitative) distinctions between things and contribute to and perhaps even *generate* these very differences. This *performative* ambiguity is captured in the phrase "the frame is involved in the evaluation

---

28. It is worth noting that Bateson writes specifically of "psychological frames," but to avoid unnecessary confusion I have left this dimension out of the current discussion.

of the messages which it contains." The last common function I want to stress is the frame's relation to communication. Bateson states that frames are by their very nature "metacommunicative": "Any message, which either explicitly or implicitly defines a frame, *ipso facto* gives the receiver instructions or aids in his attempt to understand the messages included within the frame" (188). In regard to Bateson's consideration of play, the statement "This is play" serves as an example of an explicit metacommunicative message and hence framing device: once a person states, "This is play," everything that comes after is received and responded to differently than if the statement was never uttered. Finally, the converse is also true: "Every meta-communicative or metalinguistic message defines, either explicitly or implicitly, the set of messages about which it communicates, i.e., every metacommunicative message is or de-fines a . . . frame" (188). This suggests that it is not possible to speak of something without invoking a frame, and such frames have already cut through the world before their invocation. Because a frame is defined equally by what it is not, it is not possible for a frame to be all-inclusive. One could put it as follows: *there are no frames that are open.*

Bateson's short essay was the inspiration for Erving Goffman's influential work, *Frame Analysis* (1974). In it, Goffman uses the concept of the frame to explore a basic question fundamental to all experience: How do we know what's going on in a given situation?[29] Goffman's work is an attempt to understand the "organisation of experience" and the frame forms the basis of his response. There are three aspects of Goffman's lengthy monograph that are of value in the current context:

1. Goffman's work sits interestingly in relation to debates about society and social structure, in particular the question of agency in relation to social structure. In the introduction to *Frame Analysis*, Goffman writes:

> This book is about the organization of experience—something that an individual actor can take into his mind—and not the organization of society. I make no claim whatsoever to be talking about the core matters of sociology—social organization and social structure. Those matters have been and can continue to be quite nicely studied without reference to frame at all. I am not addressing the structure of social life but the structure of experience individuals have at any moment of their social lives. I personally hold society to be first in every way and any individual's current involvements to be second; this report deals only with matters that are second. (1974, 13)

---

29. It is worth noting that Goffman follows this question with a series of lengthy qualifications, but they do not bear upon the discussion at hand.

It is a rather ambiguous passage, and one that should be taken with a grain of salt. It is Goffman's way of pre-empting any criticisms of the kind that "macro forces" are the key determinants in the shaping of life (his primary target seems to be class analysis); hence, the organization of *experience* as opposed to society. However, the question of organization, which is a close relative of structure, remains central. In ways that anticipate methodological developments in actor–network theory (ANT), Goffman's method is to begin with the situation and pose the question of organization, or indeed structure, second—but it is posed nonetheless. It is an attempt to pose the question of structure without bringing a structural answer. (Of course, one must still presume the existence of the frame before any analysis of it and therefore it would be rash to presume that this method, and those employed by ANT for that matter, can somehow bring no preconceptions at all to a given situation.) A frame is a form of structure whose existence is part and parcel of the details of the situation. Because of this, frames are somewhat fragile and penetrable; they exist through their enactment, which is always a play of repetition and difference, change and consistency. Part of the durability of a frame, however, and this leads to the next consideration, stems from its materiality.

2. Frames are not merely linguistic or communicative; they are also material. Goffman's most detailed consideration of a frame is the "theatrical frame." He conducts a meticulous analysis of how the theatrical frame marks all that takes place within it, from the different roles of people (e.g., performer, audience), differing iterative and interpretive regimes bound up with such roles (a laugh on stage may signify something different to a laugh from an audience member); different expressive and material conventions (applause, the red curtain) and different spatialities (seating area vs. stage area, backstage, cloakroom). As Michel Callon describes it in a commentary on Goffman, materiality is constitutive of the theatrical frame: "A whole series of material means are used to demarcate the theatrical space and the actions that take place within it: the building itself; its internal architecture; the bell, dimmed lights and raising of the curtain that indicates the start of the performance" (1998, 249). Writing of frames more generally, Goffman puts it thus: "activity interpreted by the application of particular rules and inducing fitting actions from the interpreter, activity, in short, that organizes matter for the interpreter, itself is located in a physical, biological, and social world" (1974, 245). Thus, paralleling Latour, we might say that although frames always have a porousness and fragility to them, *materiality is frames made durable.*

3. Even though there is a durability to frames, it is not at all the case, however, that members necessarily agree as to which frame applies to a given

situation. Responses to the question "what is going on here?" can vary greatly. Indeed, it is possible, even common, that different framing activities and different frames operate in overlapping situations. Goffman develops a nuanced language for the various occasions where there is ambiguity in play (such as keying, fabrication, misframing, and illusion) and focuses explicitly on this issue in a chapter titled "Ordinary Troubles." For my purposes, the key type of ambiguity Goffman identifies is the "frame dispute." He opens the discussion with an immediately graspable example: "It is reported that what is horseplay and larking for inner-city adolescents can be seen as vandalism and thievery by officials and victims" (1974, 321–22). Following immediately from this, Goffman defines the main features of a frame dispute:

> Now although eventually one of these sides to the argument may establish a definition that convinces the other side (or at least dominates coercive forces sufficiently to induce a show of respect), an appreciable period can elapse when there is no immediate potential agreement, when, in fact, there is no way in theory to bring everyone involved into the same frame. Under these circumstances one can expect that the parties with opposing versions of events may openly dispute with each other over how to define what has been or is happening. A frame dispute results. (Goffman 1974, 322)

Difficulty in achieving "frame alignment," coercive forces in play, open disputes—herein lies the politics of frames.[30]

Both Wikipedia Art and depictions of Muhammad raise the question of the frame. All the characteristics of framing I have described above are operative. The frame politics around Wikipedia Art exists at the level of the article entry. Rather than frame ambiguity, it seems more a question of object ambiguity: Does Wikipedia Art fit within the Wikipedia frame? But this question itself, of course, cannot be answered without making the Wikipedia frame explicit. The ambiguity of the object is at once that of the frame. While the article entry itself draws attention to the frame, this is greatly amplified during the "Article for deletion" debate. All of the policies and guidelines are *principles* for sorting. Some of the major ones mobilized in the deletion debate included: "Wikipedia:Five pillars," "Wikipedia:Deletion process," "Wikipedia:Criteria for speedy deletion," "Wikipedia:Deletion policy," "Wikipedia:No original research," "Wikipedia:Neutral point of view," and "Wikipedia:Verifiability." The "Wikipedia:Criteria for speedy deletion" policy, for example, is very clear on what lies outside the frame: "patent nonsense," "pure vandalism and blatant hoaxes," "creations by banned or blocked users," and so on.

---

30. This politicized notion of framing is what informs George Lakoff's analysis of US public discourse (2004) and Judith Butler's *Frames of War* (2009).

And even if it is not always clear when a hoax or vandalism has occurred, it is clear that when something has been identified as such it is removed. It is not merely a question of whether or not Wikipedia Art belongs in the frame, however. Framing activity is going on in several places and on different levels. The "Wikipedia:Guide to deletion" and "Afd Wikietiquette guidelines" are procedural frames. The Consensus policy frames what constitutes a settled debate. The frame sorts the outside from the inside, but also orders the inside. As the debate proceeded, frames themselves are interrogated and "higher-level" frames are brought in to settle the debate—such as when a contributor writes, "this is an encyclopedia," to frame how others should interpret Wikipedia—and these higher-level frames are themselves challenged in a search for ever-higher frames to settle the dispute. From discussion flame wars, we move to something like "frame wars."

As noted, the set of statements known as Wikipedia Art was deleted. The deletion process transformed Wikipedia Art from "encyclopedia entry" to "art stunt," or, if it was originally both of these things at once, it soon became "just art." If there was a fleeting possibility that "the Wikipedia Art page is something that explains art, explores art, and is art all at the same time," this identity was never realized—at least not in the way intended, not in the form of an encyclopedia entry. Likewise, if there was a possibility that the Wikipedia frame could be both art and encyclopedia, that the art frame and the encyclopedic frame could be made compatible, Wikipedia Art made that possibility less real, instead enforcing the noncompatibility of these higher-level frames. This sorting also had interpretive effects, which could be stated as follows: "do not approach Wikipedia Art as an encyclopedia entry; approach it as art," and conversely, "Wikipedia is an encyclopedia, which is distinct from art." Wikipedia Art was placed outside the frame, but so too were all the arguments made in favor of "keep" during the deletion discussion. Contributors such as Shmeck, Patlichty, and Artintegrated were marked as people who make invalid arguments, who don't understand the frame, while contributors like DanielRigal and Freshacconci were affirmed as productive contributors.

While the frame dispute over depictions of Muhammad did not take place at the level of the entry, it nonetheless has much in common with that of Wikipedia Art. The key framing issues in this debate are once again to do with the kind of knowledge that is compatible with the encyclopedia (e.g., religious, secular, offensive, censored, neutral) and the sorting effects this has on users and editors. To simplify a little, on the most abstract level the frame dispute was positioned as religious (Islamic) knowledge versus encyclopedic knowledge, as whether or not these frames could overlap or whether or not one excludes the other. To avoid repetition, I will only briefly highlight some of the novelties of this debate.

Because the debate is not over the article as a whole, the frame war takes place in a different setting (on the discussion pages), with different but related processes (regular discussion, mediation), and with related outcomes. The boxes at the top of the discussion page serve to frame any proceeding discussion, as well as what has come before, instructing, for example, that "the discussion of images is over" and "consensus has been reached." Along with the standard policies and guidelines, such as "Wikipedia is not censored" and NPOV, the FAQ answers emerge as organizing principles. They instruct readers that some of the crucial frame ambiguities to do with this entry have been settled and on what grounds. Entirely different mechanisms for sorting the frame also emerge, such as "protecting the article," which automatically sorts people (and their statements) based on temporality (new users) and whether or not they have a registered account. The mediation process stands in relation to "Articles for deletion" as a "procedural frame" instructing how argumentative statements are to be sorted. While the "Article for deletion" discussion ended quickly, and had a formal deadline built in (even though it was not observed), the image dispute could go on indefinitely. An unusual scenario has emerged where things marked as outside the frame are nonetheless registered inside, even if only as "requests for deletions." And even though the Muhammad dispute takes place on a lower level of the encyclopedia, the stakes are clearly much higher. The Danish cartoon controversy sits in the background, with its bombings, protests, and killings, as part of the hinterland (Law 2004). What starts out as a small dispute over an image can once again escalate to higher and higher frames, what Goffman calls "primary frameworks," marking these frameworks as incompatible along the way (as opposed to just different).

The politics of the frame is about sorting, of people and things, of statements, spaces, and regimes of interpretation, in and out, meaningful and irrelevant, and legitimate and illegitimate. Although outright frame wars are rare, there is no escaping framing, and such sorting always has political effects. A frame is always partly constituted by what it is not; it is the product of, and also produces, difference. Wikipedia is constituted by a distinct frame of knowledge, one that owes a lot to the tradition of Enlightenment. It also frames interaction; how debates can play out; what counts as agreement (i.e., consensus); how contributors' statements are to be received; who is productive, a mediator, an administrator, a troll, an artist, a radical, and who can only register inside the frame as "request for deletion." In the final section of this chapter I want explore the political limits of frame wars, especially in relation to the legitimizing function of frames and the disputes that play out inside and in between them.

Differends and Boundary Statements

What is the effect of a frame dispute? How to register the ordering quality and "political-ness" of a statement? Its micro-physics of order? Its frame affirmation and enactment? What is the threshold of legitimacy of a given dispute? To explore these questions I continue with the Muhammad dispute, considered in relation to Jean-François Lyotard's notion of the differend. Lyotard first introduced the differend in a short essay from 1982 in *Tombeau de l'intellectuel et autres papiers*, later translated and published in *Political Writings* (1993). He writes: "'Society,' as one says, is inhabited by differends. I would say that there is a differend between two parties when the 'settlement' of the conflict that opposes them appears in the idiom of one of them while the tort from which the other suffers cannot signify itself in this idiom" (1993, 9). Differends appear when there is no overarching mechanism to adjudicate a dispute that both parties recognize as legitimate. Moreover, any attempt to adjudicate, to mediate a conflict, from a "frame" that either one or both parties does not recognize is, for Lyotard, a political *wrong* (more on this in a moment). What is crucial is the kind of identity work performed by the frame and its metacommunicative framing statements: Are images of Muhammad just another educational artifact, or are they deeply offensive and idolatrous? For Lyotard this is a question of signification (and he is not wrong), but I have stressed the central role of frames as the conditions that structure regimes of signification. The following metacommunicative statement (appendix A, no. 3.5) makes the importance of the frame clear: "This is an encyclopedia, not the Quran. There is no reason not to have a picture."

On the particular *wrong* produced by an attempt to adjudicate a differend, Lyotard writes:

> Applying a single rule of judgement to both in order to settle their differend as though it was merely a litigation would wrong (at least) one of them (and both of them if neither side admits the rule). Damages result from an injury which is inflicted upon the rules of a genre of discourse but which is reparable according to those rules. A wrong results from the fact that the rules of the genre of discourse by which one judges are not those of the judged genre or genres of discourse. (1988, xi)

The *wrong* produced by the differend occurs when statements from one frame cannot be translated into another, when any attempt to include must do so in a way that has already *wronged*. Thus, while an entry on Muhammad in Wikipedia can exist, it can only exist through the lens of the Wikipedia frame and this frame produces wrongs (as well as damages) because not all conflict can be adjudicated within the principles of the frame. This inability to settle disputes amicably within any particular frame explains attempts to appeal to higher frames, where perhaps a wrong can be transformed into damages (into a frame that all parties recognize). However, as Lyotard reminds us, "a universal rule of judgement between heterogeneous genres [frames] is lacking in general" (1988, xi). There is no frame without an outside; no frame that isn't constituted by what it sorts out as well as in, and thus no escape from the politics of the frame. For Lyotard, differends are part and parcel of a world of radical, irresolvable difference.

The dispute over images of Muhammad produces a differend: those who want the images removed are not able to articulate their concerns within the Wikipedia frame. They are wronged because their knowledge statements, and in particular a nonpictorial notion of Muhammad, cannot be articulated within the Wikipedia frame. These statements do not register; they have no agency or transformative capacity and thus neither do those who express them (within this particular frame). The wrong is threefold: the object of concern (the Muhammad entry) appears in a way that is considered deeply offensive;[31] the discussion statements are deemed illegitimate; and those who express them are in turn marked as biased, religious fundamentalists, deceitful, angry and so on. In other words, there is an outcome that wrongs, a procedure, and, from these, a related set of identity effects. Lyotard's differend represents the uncrossable boundary between certain frames, their mutual exclusivity. The differend reminds us of the end point of all procedural politics and the rashness of modes of adjudication that claim to be universal and all-inclusive.

Translating the differend into this realm of micropolitical disputes and frame wars, where debates and conflicts have been followed from beginning to end or at least from beginning through to moments of intensity and back again, different aspects of this notion emerge. First, while a differend is a permanent possibility, it must be seen first and foremost as an *outcome*. A differend is the result of a dispute in which the frame and its related procedures and mechanisms of adjudication are unable to register the statements of one or more parties. To be sure, some situations are more likely to produce

---

31. Lyotard writes, "The referent is not the same when the phrase referring to it is not from the same family" (1988, 30).

differends than others,[32] but a differend is never assured. This is necessarily the case because no frame is fixed for all time. Because a differend is a contingent outcome and not a sealed destiny, it must be possible to identify its emergence, its coming into being, and the kinds of statements and enacted frames that support it.

Consider the following passage, to which I have added numbers for later reference:

> [1] Well, people who are not truthful are called "liars" in my culture. [2] And I think others would agree with me that the pictures are being deleted with untruthful edit summaries. [3] You're right, politics should not be discussed on the "Muhammad-talk" page. But I mentioned that for a sense of cultural perspective. [4] One culture believes in freedom of speech and thought, and the free flowing of ideas, on which wikipedia is based. [5] Another culture believes in something else. [6] Unfortunately, people are bringing that "something else" with them when they edit wikipedia.[33]

In addition to the existence of differends as outcomes, there are *differend-tending* statements that can be identified during a frame dispute. In the passage above, statement 1 circumscribes the existence of a culture where subjects are marked by their relation to truth. The subject uttering the statement associates him- or herself with this culture, "my culture." The interlocutor is placed outside this culture—it is not "our culture"—and implicitly accused of not being truthful and therefore positioned as a liar. Statement 2 asserts the existence of untruthful edit summaries that have been used to delete the images. That is, a process related to the referent (the images) is placed within the "culture of truth," marked as untruthful and therefore illegitimate, and in turn marks the subjects engaged in this process once more as liars. Statement 3 refers to previous discussion beyond the parameters of the extracted material, but it explicitly positions talk pages as nonpolitical, which is of course a common political gambit! Statements 4 and 5 enact a cultural divide. In addition to truth, one culture is associated with freedom—of speech, thought, and ideas—and Wikipedia is placed within this "primary" cultural frame. Statements 5 and 6 mark those who delete the images as Other, as part of an incompatible frame from the one that Wikipedia belongs to. They bring "something else" and hence *are* something else. Differend-tending statements are ones that frame a situation in a way that is not recognized by

---

32. The article for Muhammad is more likely than, say, the article for Mouse or Chess, but as I pointed out earlier, sometimes conflict is found in unlikely places.

33. *Wikipedia*, s.v. Nodekeeper, "Talk:Muhammad/Archive 9," accessed August 5, 2011, http://en.wikipedia.org/wiki/Talk:Muhammad.

one of the parties and which, if upheld within the principles of the frame, wrongs that party. They refer to the procedural moment of a frame dispute as opposed to the outcome. The above passage is a very obvious and somewhat crude case of differend-tending statements, but there are many others. The statement referred to earlier, "This is an encyclopedia, not the Quran. There is no reason not to have a picture," is also differend-tending. While not many statements actually contribute to the realization of a differend, all statements exist in relation to it, and some are more differend-tending than others.

In addition to differend-tending statements, other ordering and metacommunicative techniques emerge. This is especially the case in situations where participants sense that something like the differend is in play or, in common parlance, the deck is stacked. Some examples of different techniques can be found in appendix A. Categories 13 (Accuracy) and 15 (Quality/Relevance), for example, try to articulate, quite tactically (knowingly), a case for not including images in ways more likely to be recognized by the frame; 13.3 states, "No picture of the Prophet is accurate," and 15.3 argues that "unless the picture provides a valuable, educational purpose, it should otherwise not be included in the article." While neither of these explicitly invokes Wikipedia policy, appeals to accuracy and educational value are thought to be more frame-compatible than other argumentative techniques. Demonstrating an awareness of what is at stake, passage 11.6 adopts a quite different technique:

> Given the mayhem and deaths associated with the current backlash against the Danish cartoons (and, more importantly, their provocative reprinting), I think we can safely deduce that such depictions (of created things, but particularly of Muhammad) are seriously offensive to many mainstream Muslims—as offensive as, say, images of paedophilia. Thus, it seems reasonable to decide that the definitive wikipedia article about the prophet of Islam need not include an image of him, out of respect for the heart-felt beliefs of members of the second largest religion in the world. This is not censorship, it is consensus-building respectfulness.

This passage appeals for a *frame exception*, to allow something inside that doesn't belong (or rather, to exclude something that would otherwise belong). Interestingly, its argument about images of pedophilia is the beginning of a general principle (perhaps a counterprinciple), to the effect that there are indeed limits to the noncensorship policy; it is not a simple universal. The frame exception stands interestingly in relation to the frame: Does it change the frame? Or does its status as exception otherwise affirm the existing frame? Finally, there exists a whole series of contributions that flow away from the differend. Referring to the work of Geoffrey Bowker and Susan Leigh Star (1999), I call these *boundary statements*.

In their book about the work of classification, Bowker and Star develop the notion of the boundary object and related boundary infrastructures to refer to things that "both inhabit several communities of practice and satisfy the informational requirements of each of them" (1999, 297). Boundary objects "arise directly from the problematics created when two or more differently naturalized classification systems collide"; they are "one way that the tension between divergent viewpoints may be managed," which requires, among others things, a kind of "artful juggling" (292). Such "juggling" and "management" are crucial: a boundary object is not a mode of adjudicating, it is a "making do" in light of the impossibility of frame alignment. Its existence is multiple, even paradoxical; it is constituted by tension and hybridity. Boundary objects link frames but do not subsume them: they "are working arrangements that resolve anomalies of naturalization without imposing a naturalization of categories from one community or from an outside source of standardization" (297). Boundary objects are the practical response to the fact that differends and differend-like situations are not so uncommon.

The debate about images of Muhammad resulted in a differend, but was this inevitable? Was there a way to preserve difference, to recognize the statements of all parties as legitimate, to identify all parties and entities in ways deemed acceptable by all, that is, to turn the entry on Muhammad into a boundary object? Is it possible to satisfy all parties without gesturing to "a higher frame"? Bowker and Star note that boundary objects are "plastic enough to adapt to local needs and constraints of the several parties employing them, yet robust enough to maintain a common identity across sites. They are weakly structured in common use and become strongly structured in individual-site use" (297). The identity of a boundary object can be likened to a Venn diagram. The section where circles overlap is the common identity and the parts of the circles that do not intersect are the local, idiosyncratic aspects of the hybrid thing. If the differend-generating aspects can be pushed outside of the overlapping area and held there via "juggling" and "management," a boundary object can be established. The following passage (18.2, appendix A) is one such attempt to establish a boundary object:

> I would like to propose a solution that would minimize offence. Would it not be possible for the main article to have an icon (forgive the religious pun) indicating that a click upon it would lead to an image? There are all sorts of topics that could benefit, not just Muhammed. Many people would prefer to gain information about a distressing topic through words and not

images—car crash, for example, or starvation. Others would like to read about human anatomy or diseases without having to look at the evidence—not just genitalia, but internal organs can be considered "private parts"! Some modern art installations can fall into this category too. So if there's a fair chance of an image offending readers, why not include it on a separate page all to itself, just one click away. No censorship; lots of civility and consideration. (Archive_7)

One can see how the creation of a boundary object requires work, work that undoubtedly transforms the object in the process. The Wikipedia frame would have to allow for images to be turned into links (though still accessible) and those opposed to the use of images would have to accept the existence of images once removed from the main article page. Compromises must be made by all parties, and all frames, people, statements, and things described also undergo subtle transformation. For the boundary object to be successful, enough has to be preserved so that both parties continue to recognize the object as their own, as "in common." The above suggestion failed to establish the entry on Muhammad as a boundary object. Both parties were unwilling to accept the subtle transformations it proposed and chose instead to stress the differend-tending aspects of the entry as central to its "common identity." While differend-tending statements constitute situations in ways that enact and affirm differends, *boundary statements* move away from the differend and toward the creation of a boundary object. Figure 6 summarizes the relation between statements, frames, the differend, and the boundary object.

The unfinished circle represents the "plane of statements." It is the full spectrum of possible statements as they sit in relation to the differend and the boundary object. The bottom of the circle represents noncontroversial statements, situated well inside an agreed-upon frame. It is important to note that it isn't the substance of the statement that makes it noncontroversial, but rather that it exists within a frame that isn't contested. Disputes that emerge in this bottom area can be settled in a "just" manner, because all parties agree on the frame. In Lyotard's terminology disputes in this area create "damages," not "wrongs." Most activity in Wikipedia, and indeed in daily life, takes place in this bottom section. The level of difference enacted by any given statement increases as it moves to the top of the circle. Statements situated toward the top-left of the circle are differend-tending; they work toward the constitution of a differend. These statements enact multiple frames, but seek to "settle" disputes by either affirming one frame over another (and "sorting out" its statements) or by wrongly claiming that all statements can be dealt with under a single frame—these of course amount to the same thing. On the

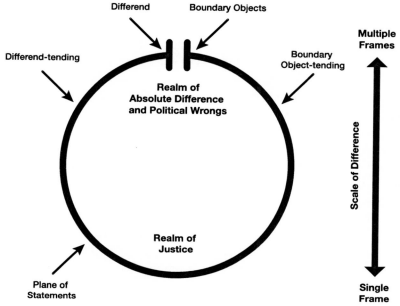

FIGURE 6. The politics of frames and statements

other side, the top-right, statements are boundary object–tending; they try to create a bridge between frames, but do not attempt to overcome the difference in any absolute way. The two lines at the top of the circle represent the extreme end points of difference. On one side lies the differend, on the other, the creation of boundary objects. The gap is pure difference. This top part of the circle cannot create "just" outcomes. At best, difference can be held at bay; at worst, it produces wrongs.

This chapter began with a consideration of collaboration as a way of understanding how people work together and how things get made within the paradigm of participation. Collaboration has been used to refer to the unique way of organizing work within open projects. We saw how collaboration was distinguished from other forms of working together, and in particular those characteristic of governments and firms operating within the conditions of the market. Although collaboration is used as a term that explains how people work together—how working together is organized—it often sits very awkwardly in relation to this very question (of organization). A whole host of terms have emerged that tend to downplay the organizing forces within collaborative work. We saw, for example, that collaboration is: "radically decentralized" (Benkler); "unmananged" and with a "spontaneous division of labour" (Shirky); self-organizing (Elliott); and that collaborative work is "non-hierarchical" and creates "*ad hoc* meritocracies" (Bruns). Without denying that such terms and related commentaries do point to genuinely novel transformations, I found them lacking in their ability to explain how an average contribution to an open project is organized. Indeed, at times collaboration appears to be a form of non- or spontaneous organization. The result of this, I suggest, is that discourses of collaboration are, like openness, depoliticized. This is perhaps not surprising, as we have had centuries to explore and feel the effects of government and market forms of organization, but very little time to experience and think about the political characteristics of open projects other than merely recognizing that they are different and seemingly better than previous forms of working together.

I shifted attention from collaboration itself to the types of ordering, sorting, and categorizing that exist within collaborative discourse, and I posed the question: if the organizational mode of collaboration is somehow obscure and fleeting, how is it that anything is produced at all?—or, how to keep out the noise? My hunch was that only organization can produce organization. In distinction to the literature on collaboration, I chose to approach the question of organization by looking in some detail at the production of two articles (and the death of one), focusing on moments where "sorting" became an issue and site of politics. Wikipedia Art and Muhammad are two very different articles, with different fates, and the types of conflict that took place on each are not reducible to one another. Moreover, the amount and intensity of conflict in these articles makes them undoubtedly exceptional, distinct to most other articles (although as I pointed out, conflict is not uncommon and can emerge in the most unexpected of places). However, rather than disregarding these articles as anomalies, I suggest that they are useful because they bring to light all the political mechanisms in Wikipedia that lie sleeping in plain sight or that are not always experienced as political because of their productive potential.

Differences and exceptionality aside, I used these two articles as the starting point for a *politics of organization* in relation to statements. This politics of organization is focused not, at least not primarily, on people, but on the organizing and sorting capacity of statements. Statements exist within and simultaneously enact frames. This can either be explicit, in the form of metacommunicative statements (e.g., "This is an encyclopedia") or implicit, where the frame remains backgrounded. While the ordering capacities of a statement might be unclear, multiple, and dynamic, statements are inherently ordered. They always exist in relation to a frame (at least one) and its principles of organization, of sorting in and out. And just as statements are ordered, they in turn order. In the Muhammad debate, I showed how statements not only competed for their own legitimacy and belonging, but how these statements also ordered certain subjectivities (such as liar or mediator), other statements, objects under consideration (the Muhammad entry), as well as the frame itself. To illustrate the political nature of frames and the "statement games" that play out in relation to them, I borrowed and reworked Lyotard's notion of the differend. Through this notion, the full stakes of a frame dispute along with the effects of being "sorted" are brought to bear. Finally, I suggested that the organizational capacities and political nature of statements exists on a spectrum, with the differend at one end and the creation of a boundary object at the other. What is crucial in this formula is that it makes no claim to be able to bring everything into the one frame. It acknowledges the radical incommensurability of certain frames and the impossibility of settling disputes in a "just" manner. Faced with the politics of the frame, sometimes the best one can hope for is the forging of a tenuous link, where otherwise there would be a differend.

Apart from showing the political dimensions of working together on Wikipedia, this new conceptual apparatus speaks directly to understandings of collaboration. The frame itself emerges as an organizing force, and this force flows over the different facets of collaborative work. While collaboration might be beyond market signals and managerial commands, as Benkler suggests, the frame has its own signals ("This is an encyclopedia") and its own principles, from which the authority of commands can be established ("This is an encyclopedia, therefore Wikipedia Art must be deleted"). The force of these commands does not stem from one's (managerial) position within a firm or other bureaucratic institution. Instead, it comes from the frame, more specifically, the ability to fit within the frame, to position oneself in relation to it, to mobilize it, and, if necessary, defer to its authority. While contributors and information architectures can accurately be described as decentralized (Benkler), contributions are nonetheless brought together and played off against one another in relation to a complex set of principles that are not weakened by decentralization. Indeed, decentralized organization can only exist if certain principles are especially forceful.

While the division of labor might not follow traditional patterns and might not be managed in terms of hierarchies of command, the frame shows that labor is not exactly spontaneous or unmanaged (Shirky). We saw that in certain instances contributions were quite predictable, determined by the constraints of the frame and the position of the contributor in relation to it. For example, during the mediation contributors were divided into those "for" and "against" the images and their contributions had to take the specific form of short position statements. Although this process broke down, contributions clearly remained divided into set categories ("for" and "against") and divided the labor accordingly. More than this, though, the frame continuously guides the process of "what needs to be done." When Shirky writes, "one person can write a new text on asphalt, fix misspellings in Pluto, and add external references for Wittgenstein in a single day" (2008, 120), it is because encyclopedias must be comprehensive, must not have spelling mistakes, and should provide references to further sources. To make clear how the frame orders work, consider if Shirky had instead written: "one person can write a second entry on asphalt, create spelling errors in Pluto, and delete valid external references for Wittgenstein." Work on Wikipedia is indeed ordered and organized in ways different to industrial or postindustrial models, but there is a logic to it.

Leading on from this, Bruns's account of work structures as nonhierarchical "*ad hoc* meritocracies" is also a bit weak. All kinds of hierarchies were created between articles, images, contributions, and contributors and some of these became stable. Bruns's "*ad hoc* meritocracies," however, refer specifically to emergent forms of *leadership* that are derived from the quality of contributions: Leaders will emerge in specific situations because the community perceives them to be the best in that instance at a particular task. No doubt leaders do emerge and hold sway over specific groups or build up authority in relation to a particular task or topic. But the nature of this leadership, let's say the source of its competence and authority, plays out in relation to the frame. As I will show in the next chapter, such sources of authority and governance, in general, are not *ad hoc*. The more a contributor masters the frame, the more likely it is that that person's contributions will be valorized within it and, in turn, that the quality of that person's contributions will increase access to positions of authority and leadership. (Of course, it is also possible to master the frame precisely to disrupt it, but that is a different issue.) We must also be very careful to qualify merit, therefore, as the "mastery of a frame," rather than as some general and absolute quality of an individual. Finally, while Reagle (and Wales) rightly point out that NPOV is a key mechanism of collaboration, I have shown that the very principles that make collaboration possible also exclude certain contributors and contributions. This is not to suggest that such exclusion is necessarily bad, just that it is necessary: the same frame that makes a coherent thing like Wikipedia possible does so by sorting out what is other. In light of claims that Wikipedia's policies provide a position from which everyone can agree and work together in harmony, even if only in theory, placing the politics of the frame—with its differends, boundary objects, wrongs, and damages—alongside collaboration is especially pressing. Indeed, it is not possible to make visible the genuinely a(nta)gonistic realities found in Wikipedia without doing so.

In this chapter I have stressed the organizational nature of Wikipedia and its political effects. While I have made reference to many policies and guidelines, and demonstrated the organizing force of statements, the focus has very much been on the *fact of organization*. I also stressed the "materiality of frames" but have said very little about what that means, especially in a formation like Wikipedia, where much of what takes place can be understood as discursive in the original Foucauldian sense. I carry these issues into the next chapter, where I look more closely at how the Wikipedia statement formation is organized. If this chapter asked, "how is it that a statement enters the formation?," the next begins with the question, "which statements attain enough force to become definitive of the formation?" In other words, what are the principles of the frame?

3

The Governance of Forceful Statements: From Ad-Hocracy to *Ex Corpore*

Bureaucracy versus Ad-Hocracy

The problem of how to organize the social, whether considered as a totality or broken down into discrete and autonomous operations, has long occupied Western thought. For a long time it went under the banner of philosophy, then sociology and political theory, and more recently it has also gone by the name of management and organization studies. In chapter 1 I briefly introduced Plato's ideas on how best to organize society, albeit read through the unsympathetic lens of Popper. Plato's *Republic* can be understood precisely as a book about organization, beginning with a discussion of its legitimating mechanisms (justice, happiness) and moving to a detailed description of Plato's ideal type of organization, aristocracy, and a consideration of its inferior alternatives. Much of the history of political philosophy can be read in this manner, as treatises on organization and its mechanisms of legitimization.[1]

During the mid- and early twentieth century it was communism, fascism, and capitalist liberal democracy that battled it out in wars of legitimacy as well as territory and blood. Meanwhile, however, the organizational form of bureaucracy quietly established itself as the preferred mode of organizing in all corners. One of the primary achievements of bureaucracy was precisely its ability to separate, at least initially, the problem of organization from that of politics and rule. Bureaucracy positioned itself as the best form of organization *to any end*. It was this achievement that made it possible for Max Weber to write: "The decisive reason for the advancement of bureaucratic organization has always been its purely technical superiority over any other form of organization" (1958, 214).

---

1. In fact, something similar has been proposed and partially carried out by Luc Boltanski and Laurent Thévenot (1999, 2006).

Of course, and often owing much to Weber's classic account, critical observers identified the many negative aspects of this form and the conditions from which it emerged. Members of the Frankfurt School, for example, followed Weber's lead of connecting bureaucracy to generalized processes of rationalization, processes that could nevertheless be put to any end. Marcuse put it like this: "The prescriptions for inhumanity and injustice are being administered by a rationally organized bureaucracy" (1991, 71). Even before Weber, though, and after being kicked out of the Communist Party, Leon Trotsky had already argued that bureaucracy was anathema to the communist destiny of the proletariat: "*Soviet bureaucracy,*" he wrote from exile, "which oppresses and robs the workers and peasants, leads the conquests of October to ruin, and is the chief obstacle on the road to the international revolution" (1954). Some years later, the Situationists would extend Trotsky's position to argue that bureaucracy had itself become "a *new exploiting class*" (cited in Knabb 1981, 256; see also Debord 1994). Bureaucracy was no longer seen as a means to be put to any end; it had been identified as an end in itself, albeit one that appeared in the guise of a means, free to recruit a whole spectrum of ends (communism, capitalism, democracy, fascism, nationalism, destiny) to serve as mechanisms for its legitimization.

The most powerful commentary on bureaucracy, however, or at least the one that continues to hold real purchase, did not focus on the class power of bureaucracy, the totalitarian and violent ends to which it was put, or on the kind of docile, "one-dimensional" subjectivities it was thought to engender. Rather, it worked against bureaucracy on its own terms, suggesting it was no longer the most "technically superior form of organization." This account was put forward by Alvin Toffler.

In his classic work *Future Shock,* Toffler (1970) wrote about the impact of "change" on society: "that roaring current of change, a current so powerful today that it overturns institutions, shifts our values and shrivels our roots." His term "future shock" describes the "shattering stress and disorientation that we induce in individuals by subjecting them to too much change in too short a time" (2):

> Future shock is no longer a distantly potential danger, but a real sickness from which increasingly large numbers already suffer. This psycho-biological condition can be described in medical and psychiatric terms. It is the disease of change. (2)

Future Shock (the book) was therefore proposed as a book about adaptation, about how to deal with change and avoid the disease of "future shock" (the condition). Toffler dedicated chapter 7 of *Future Shock* to organization, and it begins with an account of bureaucracy. Rather than dwell on the negative aspects of bureaucracy, he suggested that bureaucratic forms were themselves "groaning with change" and that we were already witnessing—circa 1970— the "breakdown of bureaucracy" (125).[2]

While acknowledging that his account was not comprehensive, Toffler identified three core features of a bureaucracy: (1) each individual "occupied a sharply defined slot in a division of labour"; (2) he or she "fit into a vertical hierarchy, a chain of command running from the boss down to the lowliest menial"; and (3) "organizational relationships . . . tended towards permanence" (126). Not surprisingly it is these three features that spelled the end of bureaucracy's supreme reign. Toffler's discussion is not at all a lengthy critique of bureaucracy, whose evils he took as a given. Rather, he described what he saw as readily identifiable empirical transformations: bureaucracy was already dying from a terminal case of future shock and in the process of being replaced. He noted the emergence of "task forces" and "projects" within the old organizational form, along with the related new position of "project manager." These were by definition temporary arrangements, designed to carry out specific tasks and then disband, and they often brought together very different forms of specialty (professions) and cut through traditional institutional hierarchies.

Such new arrangements, designed to overcome the sluggishness of the bureaucratic form, were increasingly coming to replace it. As an example, Toffler noted how after winning a military contract, one aircraft corporation "created a whole new 11,000-man organization specifically for that purpose," which included "hundreds of subcontracting firms" (132). And as a metonymic indication of the fate of bureaucratic forms, he noted how "organizational charts," with their strict hierarchical order and specified roles, were increasingly in disarray, unable to keep up with rapidly changing roles and departmental structures that were becoming commonplace in the emerging informational and "super-industrial" society (136). In short, bureaucracies were built for permanence and suited stable environments, and were thus ill-suited to the accelerated conditions of the second half of the twentieth century. Bureaucracies, and the "organizational man" who inhabited them, were highly susceptible to future shock.

2. It is worth noting that Toffler's observations about bureaucracy's demise are indebted to the writings of Warren Bennis (1965)—a debt readily acknowledged by Toffler. However, since it is Toffler's work that has had a direct impact on recent literature it will remain the focus of my attention.

Toffler coined the term "ad-hocracy" to describe the new organizational form that was transforming the old bureaucracies:

> Task forces and other *ad hoc* groups are now proliferating throughout the government and business bureaucracies, both in the United States and abroad. Transient teams, whose members come together to solve a specific problem and then separate, are particularly characteristic of science and help account for the kinetic quality of science and the scientific community. Its members are constantly on the move, organizationally, if not geographically. (134)

Ad-hocracies were by definition temporary, emerging in response to a specific problem or, as in the case of the military aircraft contract, an opportunity. They spelled "the collapse of hierarchy," a shift from vertical to "sideways" communication, and the advent of organizations of professionals, specialists and "technical co-equals"—a term Toffler borrowed from Joseph A. Raffaele. Toffler was enthusiastic about the becoming-ad-hocracy of bureaucracy because he recognized that organizational form had, for lack of a better term, *subjectivity effects*; whatever it is that constitutes and individuates a person is intimately bound up with the modes of organization that they participate in. For example, on bureaucracy Toffler writes: "Three outstanding characteristics of bureaucracy were, as we have seen, permanence, hierarchy, and a division of labour. These characteristics molded the human beings who manned the organizations" (144). In contrast, ad-hocracies demanded "a radically different constellation of human characteristics" (146). Toffler labeled this ad-hocratic subjectivity "Associative Man." I quote the following passage on Associative Man at length because it sums up Toffler's investment in change and what he saw as the major differences between the two organizational forms under consideration:

> Thus we find the emergence of a new kind of organizational man—a man who, despite his many affiliations, remains basically uncommitted to any organization. He is willing to employ his skills and creative energies to solve problems with equipment provided by the organization, and within temporary groups established by it. But he does so only so long as the problems interest him. He is committed to his own career, his own self-fulfilment.
>
> It is no accident, in light of the above, that the term "associate" seems suddenly to have become extremely popular in large organizations. We now have "associate marketing directors" and "research associates," and even government agencies are filled with "associate directors" and "associate administrators." The word associate implies co-equal, rather than subordinate, and its spreading use accurately reflects the shift from vertical and hierarchical arrangements to the new, more lateral, communication patterns.

> Where the organization man was subservient to the organization, Associative Man is almost insouciant toward it. Where the organization man was immobilized by concern for economic security, Associative Man increasingly takes it for granted. Where the organization man was fearful of risk, Associative Man welcomes it (knowing that in an affluent and fast-changing society even failure is transient). Where the organization man was hierarchy-conscious, seeking status and prestige within the organization, Associative Man seeks it without. Where the organization man filled a permanent slot, Associative Man moves from slot to slot in a complex pattern that is largely self-motivated. Where the organization man dedicated himself to the solution of routine problems according to well-defined rules, avoiding any show of unorthodoxy or creativity, Associative Man, faced by novel problems, is encouraged to innovate. Where the organization man had to subordinate his own individuality to "play ball on the team," Associative Man recognizes that the team, itself, is transient. He may subordinate his individuality for a while, under conditions of his own choosing; but it is never a permanent submergence.
>
> In all this, Associative Man bears with him a secret knowledge: the very temporariness of his relationship with organization frees him from many of the bonds that constricted his predecessor. Transience, in this sense, is liberating. (149–50)

Thus, the transition from bureaucracy to ad-hocracy, to flat, ephemeral associations of highly mobile, creative "Associative Men," committed to their "own career," was part of a narrative of liberation, underpinned by the ineluctable motor of change.

I begin this chapter by drawing attention to the history of these two competing organizational forms because they continue to deeply inform how participants and observers understand the organization and governance of Wikipedia. In particular, it is the narrative of liberation, of "bureaucracy bad, ad-hocracy good" that is carried forth into open projects. This narrative is visible in articles such as "Don't Look Now, But We've Created a Bureaucracy" (Butler, Joyce, and Pike 2008) and Nicholas Carr's writings on Wikipedia's "Emergent Bureaucracy." Carr writes:

> What a disappointing species we are. Stick us in a virgin paradise, and we create honeycombed bureaucracies, vast bramble-fields of rules and regulations, ornate politiburos filled with policymaking politicos, and, above all, tangled webs of power.... Wikipedia is beginning to look something like a post-revolutionary Bolshevik Soviet, with an inscrutable central power structure wielding control over a legion of workers. (2011, 195)

More positive accounts have coined notions like "Ad-hoc Meritocracy" (Bruns 2008a, 2008b), as we saw in the previous chapter, or "Adhocratic Governance" (Konieczny 2010), which remain very close to Toffler's original description.

Of course, these two organizational/governmental forms in no way exhaust the discussion. In fact, it has become somewhat of a pastime of Wikipedia's contributors to create highly contradictory lists of the various governmental forms thought to be operative: "Wikimedia's present power structure is a mix of anarchic, despotic, democratic, republican, meritocratic, plutocratic, technocratic, and bureaucratic elements."[3] In a very similar passage, Wales highlights that although Wikipedia is a mix of many older forms, it is not reducible to any of them:

> Wikipedia is not an anarchy, though it has anarchistic features. Wikipedia is not a democracy, though it has democratic features. Wikipedia is not an aristocracy, though it has aristocratic features. Wikipedia is not a monarchy, though it has monarchical features.[4] (2004)

And there have been a host of articles analyzing Wikipedia in relation to one model or another, from anarchy to dictatorship and everything in between.[5]

In addition to these well-worn notions are newer ones born from within the community and surrounding software cultures, such as "benevolent dictator" and "the cabal" (something approaching oligarchy), together with ones from outside commentators, such as "authorial leadership" (Reagle 2007), "peer governance" (Bauwens 2005b; Kostakis 2010), "online tribal bureaucracy" and "hacker authority"(O'Neil 2009, 2011). While commentaries alternate between considerations of governance, leadership, and authority, I consider them all part of the more general rubric of organization.

What follows, then, is not an exhaustive typology of the different modes of organization, or of whether, how, and when one particular form is in operation. Instead, my initial focus is on ad-hocracy and bureaucracy. I develop a critical account of ad-hocracy—and by extension all subsequent writings informed by it—by revisiting its relation to Weberian bureaucracy. In the

---

3. "Wikipedia Power Structure," 2011, from http://meta.wikimedia.org/wiki/Power_structure.
4. J. Wales, "Talk:Benevolent dictator," 2004, http://meta.wikimedia.org/?oldid=544462.
5. Beschastnikh, Kriplean, and McDonald 2008; Capocci et al. 2006; Descy 2006; Forte and Bruckman 2008; Holloway, Bozicevic, and Börner 2005; Reagle 2005; L. Sanger, "The Early History of Nupedia and Wikipedia: A Memoir," Slashdot, 2005, http://features.slashdot.org/story/05/04/18/164213/the-early-history-of-nupedia-and-wikipedia-a-memoir; Spek, Postma, and van den Herik 2006; Stvilia et al. 2005.

latter stages of the chapter, I further develop my own political theory of organization, building on notions introduced in previous chapters.

Why focus on ad-hocracy, given the "hybrid" nature of Wikipedia? I offer two reasons. First, as I will come to demonstrate, comparisons between ad-hocracy and bureaucracy participate in a larger narrative where the very question of organization is brought to bear. Ad-hocracy is very much the heroic David of this story, pictured as a form of fleeting or even non-organization, while the Goliath, bureaucracy, is criticized for being *too organized*, too rigid and unresponsive to its environment. The two models seem to sit at opposite ends of an organizational spectrum from which all the other forms could be placed. Presumably, self-organization, peer-governance, meritocracy, and perhaps democracy would be placed toward the ad-hocratic end, while dictatorship, monarchy, the Catholic Church, Plato's Republic, and other more "rigid" and lasting structures would lie toward bureaucracy. It also goes without saying that the ad-hocratic end of the spectrum is the "open end." In fact, in some instances ad-hocracy and openness appear interchangeable as the polar opposite of bureaucracy, such as when Carr writes: "The fate of Wikipedia—and perhaps the general 'participative' or 'open source' organizational model of online production—appears to hinge on how the tension between openness and bureaucracy plays out" (2011, 196). If we believe in this spectrum, though, the very question of organization is lost to all those things that lie at the ad-hocratic end. To pose the question of organization is always to ask *how* and not *whether or not* and, I might add, this *how* is never resolved by a mysterious "self"—as in self-governing, self-organizing, autonomous, and so forth—as if a form is completely disconnected from and unaffected by its surroundings, its ever-present hinterlands. Toffler's commentary on ad-hocracy lies at the heart of this issue.

The second reason I offer for my focus on ad-hocracy in particular is because of its function as a mechanism of legitimization. While bureaucracy was able to absorb a range of political and governmental notions to legitimate its operations, the converse is true for ad-hocracy. In Wikipedia the most contradictory forces can be mobilized—dictatorship, democracy, rough consensus, and, indeed, bureaucracy—as long as they can be legitimated by higher principles of ad-hocracy. For example, in the early years Jimmy Wales's "Benevolent Dictatorship" was thought to be tolerable because if people didn't like it they could simply leave the project and "fork" (copy the source material and start a competitor). That is, an undesirable form of governance becomes tolerable because it is coupled with a key aspect of ad-hocracy: individual mobility. So ad-hocracy is both characteristic of nonorganizational thought and, oddly enough, the organization's primary mechanism of legitimization.

Toffler's Mistake

While Toffler is celebrated as a key early interpreter of the changing organizational dynamics brought about by the morphologies of mid- to late twentieth-century capitalism (informational, networked, time-critical, and so on), his work is rarely considered in detail. Toffler's writings on bureaucracy and ad-hocracy are both complex and filled with significant ambiguities that continue to haunt the work of his contemporaries.

The first of these ambiguities is the very relationship between bureaucracy and ad-hocracy. In some sections, the relationship is one of simple succession. Toffler titles his chapter on organization "the coming ad-hocracy" and suggests, "we are witnessing not the triumph, but the breakdown of bureaucracy" and "the arrival of a new organizational system" (1970, 125). Later he writes that change and other "powerful forces . . . doom bureaucracy to destruction" (128). Change is killing bureaucracy and ad-hocracy is taking its place. Elsewhere, however, the two forms appear much harder to separate:

> Throw-away organizations, *ad hoc* teams or committees, do not necessarily replace functional structures, but they change them beyond recognition, draining them both of people and power. Today, while functional divisions continue to exist, more and more project teams, task forces and similar organizational structures spring up in their midst, then disappear. (135)

This is a more complex picture, where bureaucratic forms mobilize ad-hocracies in order to survive in conditions of rapid change. Ad-hocracies rise and fall with the currents of change, but the bureaucracy remains in the background as a constant, albeit in a radically transformed and reduced state. Ad-hocratic forms are flexible, flat, in flux, and transient, and while the forces of history (i.e., change) are pushing organizations toward this ad-hocratic form, some elements of bureaucracy must remain.

Leading on from this, the second ambiguity is whether or not ad-hocracy is or has the capacity to be an organizational form in its own right. At times, Toffler's ad-hocracy does appear to have this capacity, but in other sections (as in the passage above) ad-hocracy seems dependent on other forms and is at best a transient form. At its most extreme, ad-hocracy is a kind of anti-organization, where the poles of bureaucracy and ad-hocracy become substitutable with organization and disorganization. Consider that the core features of ad-hocracy—nonhierarchical, modular, transient, flexible—seem to connote the opposite of what is historically meant by organization—at least before swarms, hives, and other insect metaphors crept into managerial and organizational thought (Parikka 2010).

Finally, based on this blurring of ad-hocracy and disorganization, it is unclear whether Toffler is writing against bureaucracy or against organization *in general*. Note from the outset how his "Associative Man" is not defined against "Bureaucratic Man," but against "Organizational Man."[6] The associate is defined through its partial connectedness, its being in-between, always partially other to any organizational form. Toffler sways between celebrating the qualities of ad-hocracy and the associative subject it engenders, and alternatively celebrating the temporary nature of ad-hocracy. This ambiguity is captured when he writes, "where the organization man had to subordinate his own individuality to 'play ball on the team,' Associative Man recognizes that the team, itself, is transient. He may subordinate his individuality for a while, under conditions of his own choosing; but it is never a permanent submergence" (1970, 150). Is it that the team will disband, that an individual "chooses" to be on it, or the fact that the team is ad-hocratic that gives cause for praise? The fact that Toffler finishes his passage on "Associative Man" by noting that "the very temporariness of his relationships with organization frees him from any of the bonds that constricted his predecessor. Transience, in this sense, is liberating" (150) seems to suggest organization itself is the enemy of liberation.

With these ambiguities running through Toffler's work and cutting to the heart of ad-hocracy, it is no wonder that contemporary writings borrowing heavily from Toffler struggle on the question of governance and organization—that is, the question of *how*—even when they claim to be writing about these very things. When Toffler made the transition from bureaucracy to ad-hocracy, he left everyone in his wake totally confused about the relationship between organization and power. For him, the new problem was not dealing with new forms of organization, but dealing with the effects of rapid change. The promise of change had made the old power questions of bureaucratic organization irrelevant. He even went as far as making jest of the "novelists and social critics [who] are still, belatedly, hurling their rusty javelins" (142) at Weberian bureaucracy—and this despite his own profound ambiguities on the matter. Thus, rather than taking seriously the question of how ad-hocracy organizes—how, for example, are large-scale, complex forms of organization possible *despite* transience?—Toffler focused on the problems of

---

6. Granted, Toffler is also referring to William Whyte's *The Organization Man* (1956).

future shock. (It's worth mentioning here that the shadow figures of Toffler's upwardly mobile, specialist and professional Associative Man are today the multitudes of precarious workers, who may or may not be skilled, are more likely to be women, and whose "mobility" and "flexibility" are a source of exploitation, uncertainty, and hardship.)

This relative silence on the actual question of the organization of ad-hocracy stems perhaps from Toffler's idiosyncratic reading of what it is that constitutes an organization. He writes, "An organization, after all, is nothing more than a collection of human objectives, expectations, and obligations. It is, in other words, a structure of roles filled by humans" (127). He further writes that when these roles change, when they are "redefined" or "redistributed," "the old organization has died and a new one has sprung up to take its place" (127). For Toffler, the limits of an organization are the roles occupied by humans. It is not just ad-hocracies that are understood in this way. The Catholic Church, the Nazi Party, and even "coffee-break groups"—in other words, all organizations—fall under this definition. But there is much more to organization than roles—a fact well understood by Weber. As it turns out, it might be time to polish the rust off these javelins and return to Weber's initial writings on bureaucracy.

Weber opens his account by outlining six core characteristics of bureaucracies. The first reads: "There is a principle of fixed and official jurisdictional areas, which are generally ordered by rules, that is, by laws or administrative regulations" (1958, 196). What defines the bureaucracy to begin with, then, are rules, laws, and regulations. The second characteristic is very much in line with Toffler's reading and describes office hierarchy and authority, but the third takes us in a different direction. It is this third characteristic, read in light of the first and second, that I want to emphasize as the starting point for an alternative approach to organization. Weber writes:

> III. The management of the modern office is based upon written documents ("the files"), which are preserved in their original or draught form. There is, therefore, a staff of subaltern officials and scribes of all sorts. The body of officials actively engaged in a "public" office, along with the respective apparatus of material implements and the files, make up the "bureau." In private enterprise, "the bureau" is often called "the office." (1958, 197)[7]

---

[7]. Although I won't consider the other characteristics in detail, the fourth stresses the "thorough and expert training" of office management; the fifth describes how "official activity demands the full working capacity of the official, irrespective of the fact that his obligatory time in the bureau may be firmly delimited"; and the sixth connects the bureau's rules with technical learning: "The management of the office follows general rules, which are more of less stable,

What constitutes "the bureau" is not in the first instance a set of human roles, but a mix of officials, apparatuses, and files. The written documents are accorded a special position: all managerial (read: governmental) decisions must be made in relation to them. Weber depicted the document as the source of authority, the place where rules are stated and archived. Together with material apparatuses, the files, as literal documents of previous decisions, arrangements, and resulting laws, must be understood as the historical force of an organization—its mode of continuity in spite of a contingent and uncertain present. Their preservation in original condition, which bestows a relic-like character upon them, is something I will return to at the end of this chapter. It is by focusing on these three components that we might reclaim a theory of organization and governance in open projects.

"Files"

Weber's trinity of bureaucracy—the official, the apparatus, and the holy document—requires some modification and translation, but is nonetheless a good starting point for thinking about the organization of stable relations in Wikipedia. For Weber, it is written documents, the "files" preserved in original form, that manage or "govern" the bureaucratic form of organization. It is, in other words, a regime of discourse, a series of powerful statements whose function is specifically to order, that is captured by his notion of "files."[8] They are statements that define what a thing is, differentiate it from other things and organizations, settle internal ambiguities, and, most important, form the basis upon which activities and events become routines and regular proceedings.

It is deeply unfashionable to claim that Wikipedia has relatively stable and highly authoritative rules or anything approaching Weberian "files." It rubs uneasily against the laissez-faire "anyone can edit" principle and totally contradicts one of Wikipedia's "five pillars," which states that "Wikipedia does not have firm rules."[9] The subtext of this pillar reads:

> Rules in Wikipedia are not carved in stone, and their wording and interpretation are likely to change over time. The principles and spirit of Wikipedia's rules matter more than their literal wording, and sometimes improving Wiki-

---

more or less exhaustive, and which can be learned. Knowledge of these rules represents a special technical learning which the officials possess" (Weber 1958, 198).

8. For a media archaeological take on files and their central importance in "governmental practices," see Cornelia Vismann's *Files: Law and Media Technology* (2008).

9. *Wikipedia*, "Wikipedia:Five pillars."

pedia requires making an exception to a rule. Be bold (but not reckless) in updating articles and do not worry about making mistakes. Your efforts do not need to be perfect; prior versions are saved, so no damage is irreparable.[10]

And yet, this antirule sentiment exists as a *pillar* of Wikipedia, as one of five written principles that define the project. The deeply contradictory existence of this pillar is further accentuated in the second sentence: It is not the literal wording of the rule that matters, we are advised, but the greater spirit or principle. The contradiction is that one must read a literal description of the nonliteral nature of rules in order to know about said nonliteral nature!

The "Wikipedia does not have firm rules" pillar is a more formal statement of a closely related policy, titled "Ignore all rules." (In fact, the title of the pillar is hyperlinked to the "Ignore all rules" policy page, which suggests that the two are synonymous.) The policy reads: "If a rule prevents you from improving or maintaining Wikipedia, *ignore it*."[11] But despite this antirule sentiment, it is completely obvious that ignoring the rules in Wikipedia is not an effective strategy if a contributor wants his or her contribution to stick. As we saw in the previous chapter, what constitutes improvement or maintenance can be highly ambiguous and the way such ambiguities are settled is almost always precisely in relation to existing rules. This could take the form of a debate about the correct interpretation of one rule, or playing different rules against each other. As the massive discussion archives of the Muhammad entry attests, such debates are an utterly literal affair.

Leaving the "ignore all the rules" mantra aside, it is clear that Wikipedia does have firm rules—powerful and forceful statements—and this is necessarily the case. To be sure, forceful statements are not fixed for all time and may undergo various revisions and transformations; they may lose or gain force; become more or less loaded (Latour 1991); increase or decrease in precision; and angle in a different direction, making possible new associations and preventing others. Also, it is true that a new statement may come to take precedence over another. What is not possible is that there are no forceful statements, as it is these statements that constitute organization, define its parameters, make it possible, stable, and coherent, and keep entropy and noise at bay. It is these same forceful statements, their organizing capacities, that constitute the real political basis of open projects; that is why the concepts I am developing must be understood simultaneously as organizational and political.

10. Ibid.
11. *Wikipedia*, s.v. "Wikipedia:Ignore all the rules," accessed November 17, 2011, http://en.wikipedia.org/w/index.php?title=Wikipedia:Ignore_all_rules&oldid=460360231.

Wikipedia is organized first and foremost by its five pillars: (1) "Wikipedia is an online encyclopedia"; (2) "Wikipedia is written from a neutral point of view"; (3) "Wikipedia is free content that anyone can edit, use, modify, and distribute"; (4) "Editors should interact with each other in a respectful and civil manner"; and (5) "Wikipedia does not have firm rules."[12] The first specifies a genre of discourse. Its mark can be found on many aspects of the project, from its existence as a reference work and body of knowledge, to the structure of articles, their layout, and their arrangement into categories. This first pillar places Wikipedia within the long history of encyclopedias, next to the works of Pliny, Chambers, and Diderot and d'Alembert (and their small army of contributing *philosophes*), among many others. The history of the encyclopedia forms one of Wikipedia's key hinterlands, an extensive "beyond"; it is that which exists outside the Wikipedia frame but also what makes it possible. The encyclopedia pillar also "tells" people how to read the text; it instructs the reading experience. Of course, it isn't the case that a reader of Wikipedia must school him- or herself on the five pillars before being able to make sense of the work. Not at all. Rather, the statement "Wikipedia is an online encyclopedia" is itself distributed throughout the project. It is inscribed into Wikipedia's very structure, so much so that *it is Wikipedia* (i.e., Wikipedia cannot be thought without it). Its appearance as an uncontroversial pillar is the result of this distributed existence, and it is precisely this distributed nature that so strongly *frames* the project, that permits readers to identify "what's going on here."

The second pillar (neutral point of view) refers to the style and tone of content. It instructs a mode of writing, a way of translating innumerable outside events and happenings into encyclopedic artifacts. Neutral point of view instructs how to write, what forms of outside discourse can be used as source material (verifiable sources), and also how to represent a topic more generally, including which aspects to cover and how to accord weight to different positions:

> We strive for articles that document and explain the major points of view in a balanced impartial manner. We try to avoid advocacy and we characterize information and issues rather than debate them. In some areas there may be just one well-recognized point of view; in other areas we describe multiple points of view, presenting each accurately and in context, and not presenting any point of view as "the truth" or "the best view."[13]

---

12. *Wikipedia*, "Wikipedia:Five Pillars."
13. Ibid.

As the last sentence of this passage makes clear (and as I have gestured toward in previous chapters), the second pillar positions Wikipedia content in relation to the history of truth—a history that is difficult for any encyclopedia to avoid.

The third pillar (free content), quasi-legal in nature, closely aligns Wikipedia with the hinterlands of free and open source software projects considered in chapter 1. This pillar connects Wikipedia to the law, and confers upon its content a particular legal status. It inscribes a property and ownership status—"your contributions are freely licensed to the public, no editor owns any article"—and these in turn bestow a unique set of legally derived and oriented potentials—"anyone can edit, use, modify, and distribute"—on the content and its editors.[14] It is this third pillar that most distinguishes Wikipedia from its encyclopedia progenitors and it is the interplay between this pillar and the first two that generates many of Wikipedia's novelties and controversies.

The fourth pillar (respectful interaction) can be described as behavioral—encouraging respect and civility. However, it is important to note that such behavioral qualities are mainly expressed in terms of statements; that is, they are qualities of statements, not people. When the pillar instructs to "avoid personal attacks," it is demarcating certain types of statements as irrelevant. To be "well behaved" is to produce statements that stick to the topic at hand and the rules in play. Well-behaved statements are ones that "find consensus" and "avoid edit wars."[15] And when conflict is unavoidable, such statements are relegated to a suitable space, "the talk page," and follow correct procedures, "dispute resolution," until they can produce or transform into the desired, consensus-quality statements.

Five pillars, each instructing and structuring at once; *instructuring* (as in *instructus*: "arrange, inform, teach, build") different dimensions of the project, circumscribing its boundaries, and giving it a precise identity. These five pillars must be understood as the most forceful statements within the Wikipedia formation. But the pillars do not stand alone, by any means; nor have they existed "as pillars" since the beginning of the project. Before the five pillars, there was the "Trifecta," and before that a "Statement of Principles" penned by Wales and located on his user page.[16] Described on the

14. Ibid.
15. Ibid.
16. *Wikipedia*, s.v. "Wikipedia:Trifecta," accessed November 17, 2011, http://en.wikipedia.org/wiki/Wikipedia:Trifecta; *Wikipedia*, s.v. J. Wales, "User:Jimbo Wales/Statement of principles" (2001), *Wikipedia, the Free Encyclopedia*, accessed November 17, 2011, http://en.wikipedia.org/w/index.php?title=User:Jimbo_Wales/Statement_of_principles&oldid=460788947.

> **Contents** [hide]
> 1 General
> 2 Article maintenance
> 3 Article naming
> 4 Article style
> 5 Categories, lists, and tables
> 6 Content standards
> 7 Copyright and licensing
> 8 Deletion
> 9 Disambiguation
> 10 Editing practices
> 11 Editor behavior
> 12 Handling disputes and disruption
> 13 Images and other media
> 14 Notability
> 15 Project activities
> 16 Project roles
> 17 Redirects, shortcuts, and subpages
> 18 Tools and templates
> 19 User names and pages
> 20 See also

FIGURE 7. Screenshot of contents box from Wikipedia: "List of Policies and Guidelines." *Source*: http://en.wikipedia.org/w/index.php?title=Wikipedia:List_of_policies_and_guidelines&oldid=437145315 (accessed July 1, 2011).

"Wikipedia:Principles" page as "one of the oldest" statements of Wikipedia principles, these (eight) overlap considerably with the five pillars and are otherwise largely consistent with them.[17] (It is worth stressing that the page dates itself back to late 2001, which is very early in the project.) Alongside the five pillars, there are many rules, guidelines, and essays that define the project and exert more or less force on its operations (rules must be followed, whereas guidelines and essays are considered less authoritative). According to the short animated video *The State of Wikipedia*, created in part to reflect on ten years of Wikipedia, there are over 200 policies, guidelines, and essays.[18] The video further notes the existence of reviews, committees, noticeboards, Wikiprojects, and other "discussions" that may be authoritative in specific situations.

The first thing to notice about policy and guideline pages is that they are extensive, often highly repetitive, poorly ordered, and sometimes circular. Figure 7 is a screenshot of the contents box from the "Wikipedia:List of poli-

---

17. *Wikipedia*, s.v. "Wikipedia:Principles," accessed November 17, 2011, http://en.wikipedia.org/w/index.php?title=Wikipedia:Principles&oldid=447205484.

18. Jess3, "The State of Wikipedia," accessed October 10, 2011, http://www.thestateofwikipedia.com/.

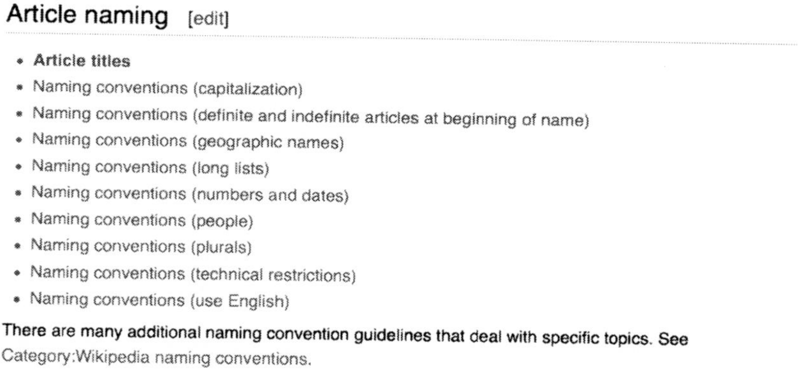

Article naming [edit]

- **Article titles**
- Naming conventions (capitalization)
- Naming conventions (definite and indefinite articles at beginning of name)
- Naming conventions (geographic names)
- Naming conventions (long lists)
- Naming conventions (numbers and dates)
- Naming conventions (people)
- Naming conventions (plurals)
- Naming conventions (technical restrictions)
- Naming conventions (use English)

There are many additional naming convention guidelines that deal with specific topics. See Category:Wikipedia naming conventions.

FIGURE 8. Screenshot of the article naming section from Wikipedia: "List of Policies and Guidelines." Source: http://en.wikipedia.org/w/index.php?title=Wikipedia:List_of_policies_and_guidelines&oldid=437145315 (accessed July 1, 2011).

cies and guidelines" page and demarcates the different aspects of the project for which policies and guidelines apply. Each of the nineteen areas is further broken down into specific sections, and these sections link to other pages where the policies and guidelines are dealt with in even more detail. There are, for example, nine general guidelines for naming articles, one policy—"Article titles," listed first and in bold—and all of which link to more detailed pages (see fig. 8). The "Article naming" section also notes the existence of "many additional naming convention guidelines that deal with specific topics." "Wikipedia:List of policies and guidelines" also links at the bottom of the first paragraph to two closely related pages, each focusing on policies and guidelines respectively. "Wikipedia:List of policies," for example, groups policies into seven areas (conduct, content, deletion, enforcement, legal, procedural, and uncategorized) and provides short summaries of each area and each policy within that area. The top of this page links to yet another page, "Category:Wikipedia Policy," which provides an alphabetically arranged list of all Wikipedia pages listed in the policy category—fifty-six in total.

Thus, along with the five pillars Wikipedia has a whole body of policies and guidelines, each of which *instructures* some aspect of the project. While there are certainly conflict-causing ambiguities within and between these policies and guidelines, ones that often play out in relation to specific developments (such as Wikipedia Art or the Muhammad images controversy), for the most part these powerful statements work to create consistency, coherency, and regularity within the project. I will not attempt to cover these comprehensively, but I will show in greater detail how one of the pillars works, its historicity, its hinterlands, how its distributed authority is exercised, and how it sits in relation to other companion statements.

The Neutral Point of View

Iterations of NPOV are coextensive with the history of Wikipedia. In his oft-cited Slashdot essay "The Early History of Nupedia and Wikipedia: A Memoir," project cofounder Larry Sanger recalls how a version of NPOV was already established in Wikipedia's precursor, Nupedia:

> Also, I am fairly sure that one of the first policies that Jimmy and I agreed upon was a "nonbias" or neutrality policy. I know I was extremely insistent upon it from the beginning, because neutrality has been a hobby-horse of mine for a very long time, and one of my guiding principles in writing "Sanger's Review." Neutrality, we agreed, required that articles should not represent any one point of view on controversial subjects, but instead fairly represent all sides.[19]

The original Nupedia policy was titled "Lack of Bias" and its core elements are strikingly similar to the current Wikipedia equivalent. Some especially pertinent excerpts include:

> Nupedia articles are, in terms of their content, to be unbiased. . . . This requires that, for each controversial view discussed, the author of an article (at a bare minimum) mention various opposing views that are taken seriously by any significant minority of experts (or concerned parties) on the subject. In longer articles, of course, opposing views will be spelled out in considerable detail. In a final version of the article, every party to the controversy in question must be able to judge that its views have been fairly presented, or as fairly as is possible in a context in which other, opposing views must also be presented as fairly as possible. . . .
>
> On any controversial issue, it is usually important to state which views, if any, are now (or were at some time) in favor and no longer in favor (among experts or some other specified group of people). But even this information can and should be imparted in such a fashion as not to imply that the majority view is correct, or even that it has any more presumption in its favor than is implied by the plain fact of its popularity.
>
> To present a subject without bias, one must pay attention not just to the matters of which views and arguments are presented, but also to their wording or the tone in which they are mentioned. Nupedia articles should avoid describing controversial views, persons, events, etc., in language that can

19. Sanger, "Early History of Nupedia and Wikipedia."

plausibly be regarded as implying some value judgment, whether positive or negative, except when the judgment is on some relatively innocuous matter and is virtually universal. It will suffice to state the relevant (agreed-upon) facts, to describe various divergent views about those facts, and then let readers make up their own minds about what the correct views are.[20]

Sanger notes how the "Lack of Bias" policy from Nupedia was translated very early on into Wikipedia as a "Rule to Consider" and shortly thereafter into the neutral point of view policy by Wales—a name disliked by Sanger for its oxymoronic status.[21] Despite the name, the core elements of the initial policy remained intact.

If Nupedia and the "Lack of Bias" policy represent the specific precursor to the set of statements that would come to be the NPOV, this initial iteration itself drew on a wider and more historical hinterland concerned with truth, knowledge, and objectivity. As Sanger notes above, he was already interested in "neutrality" before getting involved in encyclopedias and was finishing a PhD in philosophy titled "Epistemic Circularity: An Essay on the Problem of Meta-justification" (2000) when he began working for Wales. Sanger and Wales came to know one another via several mailing lists in the 1990s, although the one that stands out (at least in other commentaries) is one Wales moderated on the "Objectivist" philosophy of Ayn Rand (Reagle 2010, 57). The "Thought and Influences" section of Wales's own Wikipedia page focuses on Rand as well as Friedrich Hayek and notes that he is a "self-avowed 'Objectivist to the core.'"[22] Such a connection has led the media theorist Florian Cramer to suggest,

> Rand's "objectivism" provides the epistemological foundation of Wikipedia's open-participation authorship under a "neutral point of view." More than just a personal philosophical point of departure for the project's founders, Wales and Sanger, the idea of a world that can be generically described works as an implicit social contract binding together Wikipedia's editing community. (2011, 222)

20. *Nupedia*, s.v. Nupedia Editors, "Nupedia.com Editorial Policy Guidelines" (2000), accessed November 17, 2011, http://web.archive.org/web/20001205000200/http://www.nupedia.com/policy.shtml.

21. Sanger, "Early History of Nupedia and Wikipedia."

22. *Wikipedia*, s.v. "Jimmy Wales," accessed November 17, 2011, http://en.wikipedia.org/w/index.php?title=Jimmy_Wales&oldid=460565953. Interestingly, the *Fast Company* article offered as the source of Wales's self-avowal doesn't appear to contain that exact expression. It does note, however, that "Wales is such an objectivist that his daughter Kira, 6, was named after the heroine of Rand's first novel, *We the Living*" (Deutschman 2007).

While Wales's and to a lesser extent Sanger's history make it tempting to connect the NPOV directly with Randian objectivism, as Cramer does, there is not much about NPOV that is uniquely identifiable to Rand. The belief that it is possible to objectively describe the world—or "generically," as Cramer puts it—far exceeds her oeuvre. That said, the NPOV does have a specific orientation toward knowledge and truth, and one that indeed comes very close to what is meant by the notion of objectivity.

The NPOV page has existed since 2001, and as of October 2011 it has been edited roughly 4,500 times.[23] During this time it has undergone many transformations and the current version is much longer than the first revised version of November 2001 (the earliest version available on Wikipedia). Nonetheless, the core aspects of NPOV, its most forceful statements, remain largely unchanged. The current "Wikipedia:Neutral point of view" page has a text box near the top noting its status as an English Wikipedia policy and a "widely accepted standard that all editors should normally follow."[24] A square box to the right signals that NPOV is one of the five pillars and a "core content policy." The opening text reads:

> Editing from a neutral point of view (NPOV) means representing fairly, proportionately, and as far as possible without bias, all significant views that have been published by reliable sources. All Wikipedia articles and other encyclopedic content must be written from a neutral point of view. NPOV is a fundamental principle of Wikipedia and of other Wikimedia projects. This policy is non-negotiable and all editors and articles must follow it.
>
> "Neutral point of view" is one of Wikipedia's three core content policies. The other two are "Verifiability" and "No original research." These three core policies jointly determine the type and quality of material that is acceptable in Wikipedia articles. Because these policies work in harmony, they should not be interpreted in isolation from one another, and editors should try to familiarize themselves with all three. The principles upon which this policy is based cannot be superseded by other policies or guidelines, or by editors' consensus.[25]

The first sentence summarizes the epistemic stance of Wikipedia, with the rest of the passage indicating the force of this stance ("nonnegotiable," "can-

---

23. "Wikipedia: Neutral point of view—Article revision statistics—X!'s tools," Toolserver.org, 2011, http://toolserver.org/~soxred93/articleinfo/index.php?article=Wikipedia:Neutral_point_of_view&lang=en&wiki=wikipedia.

24. *Wikipedia*, "Wikipedia:Neutral point of view."

25. Ibid.

THE GOVERNANCE OF FORCEFUL STATEMENTS 107

not be superseded")[26] and pointing to its two key allies, "Verifiability" and "No original research." While the orientation toward bias has been slightly weakened—"as far as possible without bias"—and the concept of "reliable sources" has been included, much of this opening statement closely mirrors the first iteration from Nupedia.

What has been crucially preserved from the outset is a particular twofold relation to truth. Neutrality, defined interchangeably as nonbiased or lack of bias, attempts to distance itself from the truth-battles of the outside world, that is, contests of truth that take place outside the Wikipedia formation. For example, it no longer matters if a statement, "Jesus was resurrected," corresponds to an actual reality of the figure Jesus rising from the dead. Such distancing from these battles in turn enables an inclusiveness, where competing truths—reconfigured as conflicting "points of view"—can all be subsumed into the encyclopedic mode, albeit under quite specific conditions. For example, "Jesus was resurrected" might appear instead as "Most Christians believe in the resurrection of Jesus" or "The belief that Jesus was resurrected is a core component of Christian faith" and such statements therefore become compatible with other, non-Christian perspectives on Jesus. Indeed, the entry on Jesus contains six main "religious perspectives," a subsection on "other [religious] views," and a dedicated section on "historical views."[27]

This first relation to truth, one of distancing and inclusion, is captured in the part of the statement that requires "representing fairly . . . all significant views." It is restated in the "this page in a nutshell" box (at the top of the page) as "Articles mustn't take sides, but should explain the sides, fairly and without bias"[28] and it is equally present in the original Nupedia formulation (above). This first relation to truth also marks the limit of most commentaries on the subject. In differentiating Wikipedia from its historical counterparts, for example, Joseph Reagle writes this about the NPOV:

> Historically, reference works have made few claims about neutrality as a stance of collaboration, or as an end result. While other reference works have had contributions from thousands of people, they were still controlled by a few persons of a relatively homogenous worldview. Indeed, a preoccupation

---

26. Once more, forceful statements such as these sit uneasily against "ignore all the rules." The matter is complicated further by the weaker version of the statement in the text box above the passage: "all editors should normally follow"; *Wikipedia*, "Wikipedia:Neutral point of view."

27. *Wikipedia*, s.v. "Jesus," accessed November 17, 2011, http://en.wikipedia.org/w/index.php?title=Jesus&oldid=460999859.

28. *Wikipedia*, "Wikipedia:Neutral point of view."

of traditional references is their authoritativeness, quite different from Wikipedia's abandonment of "truth." (Reagle 2010, 56)

In Reagle's account, historical reference works are positioned as embroiled in the controversies of the day, as clearly taking an authoritative stance—that is, choosing a singular point of view—which reflects a "relatively homogenous worldview," whereas Wikipedia takes no part in these activities. It is also this first relation to truth that lends itself to enthusiastic narratives of collaboration, where NPOV is positioned as the way to overcome conflicts over competing positions. Recall the quote from Wales in the previous chapter:

> The whole concept of the neutral point of view, as I originally envisioned it, was the idea of a social concept, for helping people get along: to avoid or sidestep a lot of philosophical debates. Someone who believes that truth is socially constructed, and somebody who believes that truth is a correspondence to the facts of reality, they can still work together. (Wales, cited in Reagle 2010, 53)

In actual fact, Wales's take on Wikipedia and truth goes even further than Reagle's. It is not *particular* battles for truth that are "abandoned," but truth in general. It is this "philosophical side-stepping" that paves the way for consensus-based collaboration.

There is, however, a second relation to truth, what might be called *the truth of the NPOV* or the internal truth of the encyclopedia. I hinted at this in the previous chapter when I noted that while the NPOV doesn't claim to tell the truth about a thing, there is nonetheless a truth of what is neutral. Recall that NPOV "doesn't take sides, but should explain the sides, fairly and without bias." How does this actually work? What is the truth of this statement? Another way of putting it is: how is the first relation to truth, one of distance and inclusion, established and confirmed? The NPOV entry provides most of the answers.

After the sentence, "Observe the following principles to achieve the level of neutrality which is appropriate for an encyclopedia" is a list of five principles:

> Avoid stating opinions as facts.
> Avoid stating seriously contested assertions as facts.
> Avoid presenting uncontested assertions as mere opinion.
> Prefer non-judgemental language.
> Accurately indicate the relative prominence of opposing views.[29]

These principles spell out a precise relation to "outside" truth, at the level of individual statements. The first two require weakening the truth-value—

---

29. Ibid.

or "facticity"—of anything contested. Any statement whose truth-value is contested "should be attributed in the text to particular sources, or where justified, described as widespread views, etc."[30] The fourth principle is similar to these first two, but directs the focus to language and authorial voice. Another way of putting it would be: "avoiding adding opinions to facts."

The third principle is the converse of the first two principles: if something is uncontested, don't weaken its truth-value. This principle comes very close to participating in "truth battles" with the distinction, perhaps, that what matters is not whether or not a statement is actually true, but whether its truth is contested. The above example of Jesus' resurrection captures this distinction: The statement "Jesus was resurrected" is clearly contested and thus cannot be included. By reformulating it as "Most Christians believe in the resurrection of Jesus" the statement is weakened, at least in relation to the resurrection of Jesus. The focus of the truth-value of the statement has actually been redirected from the resurrection of Jesus to whether or not most Christians believe this to be true. Because this element of the statement, "Most Christians believe," is an "uncontested assertion," it therefore cannot be presented in a weakened form: "It is the opinion of X that most Christians believe...." And despite Wikipedia's "abandonment" of truth, it nonetheless has a whole regime (of truth) in place for determining whether or not this transformed statement, with its altered focus on "Most Christians believe," is in fact uncontested. The final principle requires all of these newly formed statements to be ordered in relation to one another and this order is determined by an outside reality.

Together, these five principles explain the how NPOV is established at the level of individual written statements, but by no means do they represent the limits of this pillar. Neutrality procedures also apply to the naming of articles, the structure and arrangement of articles, the "weight" given to particular perspectives (e.g., 1,000 words on a minor perspective, while the majority view is only 100 words long), research methods for acquiring sources, and so on.

There is, therefore, a whole other relation to truth to be found in the NPOV pillar. The truth-value of a statement is not at all rejected, just redirected. And while I have described this regime of truth as the internal truth of Wikipedia, in actual fact the twofold relation to truth cannot be grasped entirely in terms of an inside (a truth of NPOV) and outside (the truth battles beyond the encyclopedia). Instead, the reach of NPOV extends well beyond the limits of the encyclopedia. NPOV must be understood as a grid of

30. Ibid.

intelligibility, a set of forceful statements that circumscribe a world beyond the encyclopedia as well as the precise manner for how to engage with it; it is an internal truth with an external reach. It is also at this point that the NPOV's policy allies become especially important.

Alongside NPOV, "No original research" and "Verifiability" make up Wikipedia's three core content policies, which are designed to work in unison: "Because these policies work in harmony, they should not be interpreted in isolation from one another."[31] "No original research" establishes a preexisting outside world as the only legitimate source of encyclopedia statements. But the *outside* invoked by "No original research" is very specific: "Wikipedia does not publish original thought: all material in Wikipedia must be attributable to a reliable, published source."[32] The pre-existing outside world is purely discursive, a world comprised solely of sources. It is on this level, or in regard to this outside that Wikipedia engages in battles for truth. While I won't go into detail, there are extensive criteria for what constitutes a reliable source, a published source, and, indeed, a source in itself. The function of "Verifiability" in this regard is to establish the reality of this outside world of sources and the method for connecting to it (via citation).

NPOV is the pillar of all content policy, working in "harmony" with "No original research" and "Verifiability." Together, these three core content policies sit atop a whole body of related policies, guidelines, and essays, which all work to define the contours of the project: the precise rules of a statement's formation and the threshold of statement inclusion; the arrangement and relation between statements; and what constitutes the "source" world beyond the encyclopedia formation and how to approach it. While outside battles for truth are explicitly rejected—"The threshold for inclusion in Wikipedia is *verifiability, not truth*"[33]—Wikipedia nonetheless has a whole body of forceful statements whose function is to establish the truth of any particular statement; a truth of what is neutral, (non-)original, published, reliable, attributable, and verifiable. It is this body of written rules, the work they do to define the limits and correct procedures of Wikipedia, their position as source of authority, as the base from which the project can be "managed," which corresponds to the "files" of Weber's bureaucracy.

31. Ibid.
32. *Wikipedia*, s.v. "Wikipedia:No original research," accessed November 17, 2011, http://en.wikipedia.org/w/index.php?title=Wikipedia:No_original_research&oldid=461002085.
33. *Wikipedia*, s.v. "Wikipedia:Verifiability," accessed November 17, 2011, http://en.wikipedia.org/w/index.php?title=Wikipedia:Verifiability&oldid=461059539.

"Apparatus of Material Implements"

In his account of bureaucracy, Weber wrote very little about the precise nature of the apparatus of material implements. His approach was more historical and comparative, leaving many of the working details of this organizational structure aside. An inclusive definition of the apparatus would account for everything that constitutes the bureau or office, minus the files and officials. Presumably, this would begin with the architecture and basic infrastructure of the office—its size, location, spatial arrangement, lighting, building material, energy source, and so on. The origins of these basic components form the hinterlands of the materiality of the office; assembled in one place they represent the material conditions of possibility from which a specific organizational form emerges. The apparatus would also include everything within these basic infrastructures, from furniture and other office equipment, to minuscule things like writing utensils.

Identifying the spectrum of implements that constitute the materiality of Wikipedia is not an easy task. It obviously includes the project's many servers, various forms of software and code of which MediaWiki is probably the most important, as well as the physical space of the Wikimedia Foundation and national Wikipedia chapters. But perhaps it also includes the personal computers, mobile devices, screens, keyboards, navigation devices, operating systems, and web browsers that are the necessary conditions for contributing to the project. And what of the fiber-optic and copper cables, wireless networks, Internet protocols, and indeed energy grids required to make these other things function? I make no attempt to sort out these tricky questions or to distinguish where the apparatus ends and its hinterlands begin. All that is required is the acknowledgment that Wikipedia has this dimension. Nor will my account of the apparatus be comprehensive. Instead, what I want to capture is something of the governing capacity of the apparatus. To do so, I draw from one specific example: HagermanBot.[34]

Bots are scripts—small computer programs—designed to perform automated tasks without ongoing human assistance. As of 2008, there were 457 different bots that had recorded at least one edit on Wikipedia; 202 of these had made an edit within the past month and were thus considered active. Of these 202 active bots, 123 had made over 250 edits in the last month, and 17 had made over 10,000. A total of 39 bots had made over 100,000 edits

---

34. The example of HagermanBot is taken from the valuable work of R. Stuart Geiger (2011).

over the life of the project.³⁵ Seventeen of Wikipedia's twenty most prolific editors are bots and collectively bots make around 16 percent of total contributions to the English language Wikipedia (Geiger 2011, 79). Other language versions have a much higher percentage of bot activity (Niederer and van Dijck 2010, 14–15). Bots perform any number of tasks that lend themselves to routinization, from scraping and adding web content (such as demographics from government and related sites) to enforcing behavioral policies and fighting vandals.

This presence of bots has been used by Geiger (2009, 2010, 2011) and Niederer and van Dijck (2010) to refute the popular "wisdom of the crowds" thesis (Kittur et al. 2007; Kittur and Kraut 2008; Surowiecki 2004) and other related positions.³⁶ For these authors the "wisdom of the crowds" crucially overlooks "the hidden order of Wikipedia" (Geiger 2011, 80) and its increasing "technicity of content" (Niederer and van Dijck 2010). Their argument, in short, is that Wikipedia's coherent order and relative stability are not the miraculous result of "many minds" working in harmony, but of sophisticated technical actors that add, filter, monitor, revert, guide and shape new contributions. Beyond this basic critique, Geiger's work in particular details what might be called a bot micropolitics. His work explores how the automation of specific tasks transformed the political fabric of Wikipedia, generated new controversies, and led to new sociotechnical arrangements based on novel compromises. In what follows, I draw on and extend his detailed account of HagermanBot as the basis for my own thinking on the Weberian apparatus.

HagermanBot was retired on May 25, 2007, after roughly six months of diligent service, and replaced by the more sophisticated SineBot. The main function of these bots is to append information to unsigned comments in the talk pages of Wikipedia. Without going into too much detail, wikis are not as well designed for discussion as other software architectures. The fact that contributors are able not only to add their own comments to a discussion page, but also to edit previous comments (including the comments of others), is just one obvious limitation. To overcome these wiki-software issues a series of norms emerged to create a more coherent discussion page structure,

---

35. *Wikipedia*, s.v. "Wikipedia:Editing frequency/All bots," accessed November 17, 2011, http://en.wikipedia.org/w/index.php?title=Wikipedia:Editing_frequency/All_bots&oldid=254264333.

36. For Niederer and van Dijck, related concepts include: "many minds" (Sunstein 2006), "distributed collaboration" (Shirky 2008), "mass collaboration" (Tapscott and Williams 2006), "produsage" (Bruns 2008a, 2008b), "crowdsourcing" (Howe 2006), "open source intelligence" (Stalder and Hirsch 2002), and "collaborative knowledge" (Poe 2006).

one of which was to sign and date new comments. This practice would help contributors keep track of who wrote what and when, which can be very important when a discussion page is large and very active. HagermanBot functioned by patrolling the "recent changes" log on Wikipedia and appending an "Unsigned template" to any new talk page contribution that satisfied the now-technical quality of being "unsigned." Templates are chunks of information that can be linked to by other pages using the Wiki markup template syntax. Mostly, templates add generic or highly repetitive information to pages and can be used by humans and bots alike. For example, the Unsigned template is written as follows:

{{ subst:unsigned | user name or IP | time, day month year (UTC) }}

The double curly brackets flag the contained text as a template. The Unsigned template specifically gathers username or IP address details (for contributors without user accounts) and date information from the talk page edit history log. After this process is complete, the template produces text such as:

Preceding unsigned comment added by LarryW (talk • contribs) 11:03, 1 June 2011 (UTC)

Or for anonymous contributions, where the username is replaced with an IP address and the link to the users' previous contributions (contribs) is removed:

Preceding unsigned comment added by 255.255.255.255 (talk) 11:03, 1 June 2011 (UTC)

As a curious anthropomorphic side effect of the MediaWiki software, the procedure for bot activation requires a bot to register a user account of the same type as human users. On behalf of HagermanBot, user Hagerman (the bot's creator) registered a new account on November 30, 2006 (Geiger 2011, 84). However, new bot users must also be approved by the Bot Approval Group (BAG) before they can begin operating. The BAG ensures bots adhere to Wikipedia's Bot Policy, which in turn requires that bots adhere to all other Wikipedia policies. In order to be approved, Wikipedia's Bot Policy requires that a bot:

- is harmless
- is useful
- does not consume resources unnecessarily
- performs only tasks for which there is consensus
- carefully adheres to relevant policies and guidelines

- uses informative messages, appropriately worded, in any edit summaries or messages left for users.³⁷

HagermanBot was quickly approved by a member of the BAG, Tawker, and began operating on December 3, 2006. According to Geiger (2011, 85), HabermanBot autosigned 790 comments on the first day, and more than 5,000 over the next five days. It was enough for HagermanBot to outedit all human editors and almost all other bots.

Geiger tells the story of HagermanBot in the language of actor–network theory. For Geiger, the bot is an instance of "delegation," where the formerly tedious procedure of patrolling logs and attaching the Unsigned template to unsigned contributions is replaced by technical means. Drawing on Latour, he notes that it is not the quality "human" or "nonhuman" that is of primary interest, but the effects of substituting one actor with another. It is via the substitution of humans by HagermanBot, for instance, that the notion of "unsigned" becomes a technical condition—a quality or nonquality of a group of words in the presence of code. Importantly, it is not only the actual act of attaching the template that is automated, but also the related decision-making process: HagermanBot must define what is signed and what is not.

By autosigning all unsigned talk-page contributions, HagermanBot relieved other editors from the dull task of attaching the Unsigned template and simultaneously created more order in the discussion space, thereby making discussion easier to follow. Through the activation of HagermanBot, one set of signing procedures was replaced by another. Various, noncontinuous human monitors were replaced with always-on software, ensuring that nothing would be missed by lack of attention. At the same time, however, the kind of attention exercised by HagermanBot was different to that of human editors. As noted above, "unsigned" becomes defined through technical means, specifically as the absence of four tildes (~~~~) at the end of a contribution. Four tildes is a piece of wiki markup language that functions in a similar way to a template in that it automatically signs and date-stamps a contribution. If a user fails to attach this markup, HagermanBot springs to action.³⁸

Geiger likens the operations of HagermanBot to Latour's account of the speed bump (in French they are called "sleeping police officers"), where a road rule—that is, a statement about the required speed of vehicles—be-

---

37. *Wikipedia*, s.v. "Wikipedia:Bot policy," accessed November 18, 2011, http://en.wikipedia.org/w/index.php?title=Wikipedia:Bot_policy&oldid=455017920.

38. I should note that there are other ways to sign a post that is recognized by bots, such as attaching three or five tildes, which offer different kinds of signatures.

comes infrastructure and thus exerts a ruthless morality on hoons (young men who drive recklessly) and ambulances alike (Latour 1992, 244; 1999, 187). The statement "slow down here," which is itself the product of a complex set of relations between "engineers and chancellors and law makers" and caught up in a discourse of statements about the safety of populations, is *loaded* with "gravel, concrete, paint, and standard calculations" (1999, 190). Thus, while all (written) statements are material in the sense of having an empirical existence—ink on paper, for example—Latour's version of statements, developed in detail in his essay "Technology Is Society Made Durable" (1991) among other places, offers a greatly expanded sense of materiality—one that actually breaks down the Foucauldian distinction between the discursive and the nondiscursive. Latour's statements might have linguistic components, but they cannot be isolated to the realm of discourse. In discussing his famous example of the hotel manager who wants customers to leave their keys at the hotel reception when departing, Latour notes how statements can be loaded any number of ways:

> The grammatical imperative acts as a first load—"leave your keys"; the inscription on the sign is a second load; the polite word "please," added to the imperative to win the good graces of the customer constitutes a third; the mass of the metal weight adds a fourth. The number of loads that one needs to attach to the statement depends on the customers' resistance, their carelessness, their savagery, and their mood. It also depends on how badly the hotel manager wants to control his customers. (Latour 1991, 105)

With each loading, the statement "leave your keys" is transformed into something (materially) different but it nonetheless retains important continuities (of function, for example). It is defined less by its materiality than by its trajectory of force and its ordering function. Latour continues:

> By statement we mean anything that is thrown, sent, or delegated by an enunciator.... Sometimes [statements refer] to a word, sometimes to a sentence, sometimes to an object, sometimes to an apparatus, and sometimes to an institution. In our example, the statement can refer to a sentence uttered by the hotel manager—but it also refers to a material apparatus which forces customers to leave their keys at the front desk. The word 'statement' therefore refers not to linguistics, but to the gradient that carries us from words to things and from things to words. (106)

Through Latour's unique take on statements, it is possible to understand the rise of HagermanBot as the becoming-technical of the statement "sign and date your comments." It is a loading and a transformation that results in the statement becoming part of the Wikipedia apparatus.

But loading this statement via technical means and thus making it more forceful also changed the entire dynamics of signing. HagermanBot's "ruthless morality" and rigid technical definition of what constitutes "unsigned" was challenged by some disgruntled users. They complained that signing comments was not a strict policy, or that HagermanBot was too quick to sign a comment (where perhaps an editor intended to make comments over two separate sessions and sign only once at the end). One editor complained that while they signed all their comments, they preferred not to use the automated four-tilde approach and because of this HagermanBot was double-signing their comments (Geiger 2011, 85). As Geiger points out, editors also felt embarrassed because it appeared as though they were deliberately eschewing signing guidelines; it sorted them in undesirable ways.

The activities of HagermanBot created a significant controversy, powerful enough to force change on the newly emerged post-bot dynamics of signing. While HagermanBot was allowed to continue patrolling unsigned comments, several other technical implements were developed to help settle the controversy. Those who didn't want HagermanBot stepping in could add themselves to an opt-out list or add a "disable HagermanBot" tag to their account. This in turn set a precedent for all future bot protocol and the creation of a generic "no bot" template (written as {{nobots}}). Along with this template emerged a new technical category, "Exclusion Compliant," which refers to whether or not a bot can be disabled (Geiger 2011, 89).

Apart from serving as a counterpoint to the "wisdom of crowds" thesis, Geiger uses the story of HagermanBot to capture the becoming-visible of technological actors on Wikipedia and their participation in the procedures of governance. The complex figures of bots "are both editors and software, social and technical, discursive and material, as well as assembled and autonomous." Bots are also evidence of the fact that "who or what is in control of Wikipedia is far less interesting than the question of how control operates across a diverse and multifaceted sociotechnical environment" (92). HagermanBot increased the force of certain pre-existing norms about signing comments and in doing so made their exceptions highly visible and therefore vocal. In order to become a permanent actor on Wikipedia, HagermanBot had to become more discriminant; it had to allow people to escape its grasp. Exception Compliance, via various opt-out mechanisms, represents this increased discrimination between users. In terms of the force of the bot's activity, it was both strengthened and weakened by Exception Compliance. On the one hand, the force of the (now-transformed) statement "append the Unsigned template to comments that do not include four tildes" was weakened by the addition of an opt-out exception, which limited the domain of the bot's influence. On the other hand, this act of weakening also guaranteed the bot's continued existence. It demonstrated that the bot was "well behaved," that its force could be nullified, and because users *could* opt out they were actually far less likely to.

In relation to Weber's apparatus, HagermanBot shows the birth of a new material implement; it is the becoming-technical of a previous guideline. It shows how material implements distribute different kinds of agency, experienced alternatively as increased efficiency and talk-page readability or as the imposition of clumsy signature policing and being marked as an outsider. What is a simple instrumental tool for some is a frustrating obstacle for others. HagermanBot also demonstrates the overlapping of the realms of written documents and the apparatus. Forceful statements flow back and forth and are transformed in the process, generating new dynamics within the larger formation. Indeed, while I have kept Weber's distinction between written documents and material implements, whether a statement fits into one these categories is secondary to the consideration of its force and capacity to organize.

Following the *introduction* of HagermanBot also allows us to see the functioning of this material implement in a specific way: When a new apparatus-thing is introduced, its transformative effect is also directly registered; we are able to see exactly how HagermanBot governs the procedure of signing comments and how it differs from the previous mode. HagermanBot presents as a *governing-thing*, as a participant in the governance of an organization. It is a statement-transformation that momentarily interrupts and thus makes visible the ordering principles of the Wikipedia frame. And while this implement performs only minuscule tasks, upholding minor behavioral guidelines on pages that aren't even immediately visible to most visitors, the effects of its governing capacity nonetheless rippled through the project.

Finally, HagermanBot provides insight into the performative character of statements. The force of a statement is defined primarily in relation to its performativity, in its ability to exercise itself and withstand "trials of strength," as Latour puts it (1987). By automating the process of signing unsigned comments, and by continuously patrolling Wikipedia for these occurrences and responding to them, HagermanBot increased the performativity of the statement "comments should be signed and date stamped" and therefore increased its force (and this remains true in spite of the opt-out option, to which relatively few people opted in). Every time a statement is deployed and goes unchallenged it increases in strength, shaping the formation in its own image. When a statement is challenged and is able to overcome the challenge, its strength increases even more (despite any transformations that result from the challenge).

HagermanBot, it seems, is especially suited to a description of performativity. After all, the story is one of a loose behavioral guideline becoming an automated bot. In a *Tron*-like fantasy one can imagine little robots patrolling talk pages and logs, looking for rogue comments and branding them with a magic Unsigned wand. And it is the seemingly autonomous and active nature of software that informs Hayles's (2005, 50) and Galloway's (2004, 166) position that code is performative in a much stronger sense than other kinds of language. Hayles writes,

> Code that runs on a machine is performative in a much stronger sense than that attributed to language. When language is said to be performative, the kinds of actions it "performs" happen in the minds of humans, as when someone says "I declare this legislative session open" or "I pronounce you husband and wife." Granted, these changes in minds can and do result in behavioral effects, but the performative force of language is nonetheless tied to the external changes through complex chains of mediation. (2005, 50)

The story of HagermanBot is indeed one of increased performativity and this increased performativity *is* related to the statement's translation from traditional language (in the form of a guideline) to code—it is "loaded" with code—but the mistake that Hayles and Galloway make is to ascribe a performative strength to code *before the fact.* The performativity of a statement, whether manifested as written or spoken language, code, a weight added to a hotel key, or a concrete speed bump, must be determined empirically after the fact. While Hayles recognizes that the performative force of language is dependent on "complex chains of mediation," she fails to recognize that the same is true for code, for all statements. If HagermanBot was rejected by the BAG, or if every single editor on Wikipedia joined the opt-out list, the performative force of HagermanBot would be minimal. The performative force of a speed bump on an abandoned road is equally zero, while the force of a spoken or written command that everyone obeys without question is very high.[39]

---

39. Indeed, it is surprising that Hayles and Galloway, who are both well versed in philosophy and literary theory, consider performativity largely only in terms of Austin's speech acts, without considering Foucauldian or Deleuzian perspectives on the performativity of language (i.e., as statements or order-words).

HagermanBot is a minor implement, a tiny actor within the apparatus. Through this bot's creation, guidelines regarding the signing of comments are inscribed into the apparatus. The governing role of the apparatus is increased and the organization of Wikipedia overall is transformed. The contested nature of this transformation, its felt impact on editors, made the governing component of HagermanBot highly visible—one can grasp its performative force in operation. But the automated figure of the bot shouldn't mislead. The bot doesn't act alongside other active humans on an otherwise docile architecture. The exact opposite is the case. HagermanBot draws our attention to the performative force of the entire apparatus. Most of the time this force is so inscribed, its existence enacted time and time again and its previous controversies so long ago settled, that it goes unnoticed. We no longer notice how the MediaWiki software cuts article entries into project, discussion and history sections, how the written component of pages is structured or what semi-permanent fixtures appear in the left column (indeed, we rarely consider the existence of this column itself)—a whole gamut of things, in other words, belonging to the question of generic design and which enact "high-level epistemic assumptions about how an encyclopedia ought to be constructed" (Geiger 2011, 79). And all this still belongs to the realm of software and says nothing of the list of other infrastructural considerations I put forward at the beginning of this section. It is to the entirety of these governing forces that Weber's concept of the apparatus of material implements draws attention.

"Body of Officials"

What to make of Weber's body of officials, his "staff of subaltern officials and scribes of all sorts" (Weber 1958, 197)? While it is not possible to develop a full theory of agency in relation to governance, I want to use this final section to make some general observations on these difficult matters—that is, on the governing role and capacity of humans in organizations, on their capacity for *authority*. Thus far, it might seem as though there is no room for this third member of the trilogy of governance, that governance is *what happens* to people in an organization, but the story is more complicated. In fact, it is already hinted at in Weber's primary distinction between "subaltern officials and scribes": on the one hand, a position "underneath," one who carries out tasks cast down from "above"; on the other, the *in*scriber, who acts upon the holy documents and who creates the implements that become the "above" of others.

Wikipedia has a hierarchy of human roles that are defined in terms of access and permission. While there are a range of specialist user-access levels,[40] the basic structure, in order of least to most access, is as follows: blocked user, unregistered user, new user, autoconfirmed user, administrator (sysop), bureaucrat, and steward. Other than blocked users, all users can edit pages. However, unregistered and new users cannot edit semi-protected pages (or controversial pages) or upload files (such as images). Autoconfirmed users are ones whose account is at least four days old and with at least ten edits. These users can upload files, edit semi-protected pages, and also move pages (if they have been renamed, for example). Beyond these basic access levels there is quite a leap. The access level of administrator and everything above must be granted by the community. Bureaucrats can add and remove administrators and add (but not remove) other bureaucrats. Stewards can add and remove both administrators and bureaucrats, and the role of steward is attained via election. Administrators, bureaucrats, and stewards can all edit fully protected pages (such as the front page and highly controversial pages); delete and protect pages; block and unblock users; and ascribe certain permissions to specific users (such as rollback or ipblock-exempt rights). As users participate more in the project (in ways considered constructive), they gain the ability to edit controversial content. From administrator and above, however, users gain not only access to new editing tools and permission to edit controversial content, but also the ability to act on other users (e.g., block, unblock, promote). I must stress, however, that this is a very basic sketch of what is in reality a much more nuanced hierarchy.

Without denying the significance of the hierarchy, whose specificity I will return to below, I want to suggest that the governing role of users does not derive from these roles per se. Roles based on access and permission are not the primary manifestation of human authority. Or, more precisely, these roles provide a very specific set of capacities that capture only a very small segment of the overall governance of the project. What is most important is that increased access and permission do not significantly transform a user's relation to the project's framing statements: "administrators should be a part of the community like other editors, with no special powers or privileges when acting as editors," and further, "Administrators are expected to follow Wikipedia policies."[41] To stress the nonauthoritative nature of these roles,

---

40. Of particular note are oversight and checkuser access, which sit between bureaucrat and steward. See the page "Wikipedia:User access levels" for more details.

41. *Wikipedia*, s.v. "Wikipedia:Administrators," accessed November 21, 2011, http://en.wikipedia.org/w/index.php?title=Wikipedia:Administrators&oldid=461559841.

they are often compared to the role of a janitor and the special editorial tools that comes with the role to a janitor's mop.[42] And as far back as early 2003, Wales wrote:

> I just wanted to say that becoming a sysop [administrator] is *not a big deal*.... I want to dispel the aura of "authority" around the position. It's merely a technical matter that the powers given to sysops are not given out to everyone.
>
> I don't like that there's the apparent feeling here that being granted sysop status is a really special thing.[43]

These roles are in fact designed to hold as little authority as possible and not to grant the position holder any special privileges in the face of the frame. Consider the increased access and permissions granted to an administrator: administrators can delete and protect pages, but they cannot simply delete any page at whim. Rather, there is a lengthy set of procedures—such as "articles for deletion," covered in the previous chapter—for determining when it is appropriate to delete a page and there are similar conditions for the process of protecting a page.[44] Administrators can also block and unblock users but once again there are specific sets of statements that describe when such action is appropriate and how it must be done.[45] An administrator cannot simply block someone because the administrator doesn't like the contribution: the value of a user's contribution is not determined by the authority of a rogue individual, but by whether or not the contribution satisfies the conditions of the frame. Is it neutral, verifiable, nonoriginal? and so on. Thus, holders of these roles do not have any special privileges in relation to Wikipedia's core policies, and the increased capacity to act on other users is itself subject to an equally large body of rules.

The authority of a user lies instead in his or her ability to act *in unison* with the project's framing statements and in the ability to mobilize the relevant rule or guideline whenever that user's contributions are challenged. We saw in the previous chapter, for example, how both controversies (Wikipedia Art and Images of Muhammad) were settled precisely by participants playing different rules off against one another. From this perspective, an

---

42. *Wikipedia*, s.v. "User:Hiding/Admin standards," accessed November 21, 2011, http://en.wikipedia.org/w/index.php?title=User:Hiding/Admin_standards&oldid=306024527.

43. J. Wales, "Sysop status," 2011, http://lists.wikimedia.org/pipermail/wikien-1/2003-February/001149.html.

44. See, for example, "Wikipedia:Deletion policy" and "Wikipedia:Protection policy."

45. *Wikipedia*, s.v. "Wikipedia:Blocking policy," accessed November 22, 2011, http://en.wikipedia.org/w/index.php?title=Wikipedia:Blocking_policy&oldid=458706425.

administrator might still be more capable of acting authoritatively because it is very likely that the administrator is more familiar with the project's framing statements, but this is a consequence of time and not position. Indeed, the more authority a user attains, the more a project's organizing statements flow through the user's expressions, in terms of the user's ability to explicitly invoke such framing statements when required, but also, and most important, in the user's ability to act in accordance with the frame on an ongoing basis. I call this governing capacity of users *instructuring agency*. Another way of describing it is *becoming a Wikipedian*.

Instructuring agency captures follow- or enact-the-rules-type agency, but users clearly don't always follow the rules; sometimes they change them. This distinction between following the rules and creating the rules has been discussed in a lecture by Latour titled "What's Organizing? A Meditation on the Bust of Emilio Bootme in Praise of Jim Taylor" (2008). In this lecture, Latour reflects on his recent appointment as a "rather inefficient and powerless Dean." He describes a situation where he and some fellow administrators have to make a decision about their organization (a university) in the face of uncertainty. It is a moment where the "essence of the organization," that thing that would make clear exactly what the organization should do, is not at all obvious. In considering this moment, Latour refers to the notion of a "script," which is something very similar to what I have referred to as framing statements and also, perhaps, to what Weber describes as files:

> we live at different moments under and at some times above what I would call a script. A script is a set of goal-oriented instructions that delegate to some other actors more or less specific tasks, depending on those actors' competence. A script also includes, and this is a crucial feature, deadlines, at which it will be reconsidered. (2008)

For Latour, what is crucial in the notion of a script is the differing relations people have with it over time and in different situations: "What counts for now is not so much the literary connotation of the word 'script' as the peculiar situation of finding oneself either above or below the same set of instructions." In this sense, agency—considered here as the ability to act *on* a script as opposed to *with* it—is distributed differently in regard to different people and even the same person at different times. Deadlines are important because they represent the moment where certain members of an organization go from acting underneath the script to acting above it, from following the script to revising it. The board meeting Latour found himself in, deciding on the future of the organization in the face of uncertainty, was one such moment of acting above the script. With Wikipedia, there are no such deadlines, although there are regular moments of uncertainty. Instead, a key moment of acting above the script occurs when users rewrite policies and guidelines.[46]

Another important component of the script is its indeterminacy: "most if not all of the script will be at worst contradictory and at best ambiguous or incomplete. Remember Wittgenstein's demonstration that it's impossible to make a rule completely explicit" (2008). It is not as if the script (the frame) is perfectly clear at one moment, then acted on and transformed into a revised but equally clear state. It is the very ambiguity of the script that demands it be acted on, that demands this *transformative* type of agency. I also want to suggest, by way of extending Latour's ideas, that it isn't only these exceptional moments—when the "essence" of a rule is unclear—that demand *transformative agency*. It is equally present in the ambiguity created *between* rules. Should I "demonstrate good faith" (a behavioral guideline) by compromising my desire to have images of Muhammad or should I include images because "Wikipedia is not censored" (a content policy)? Conversely, even though one editor has included images of Muhammad and states she is entitled to do so (because Wikipedia is not censored), should I "ignore the rules" (a pillar) and remove the offensive images? If it is impossible to make a rule completely explicit, it is equally impossible to create a perfectly clear relations between rules.

Finally, although Latour stresses the importance of deadlines as instances of being above the script, further into his lecture he acknowledges that, "in practice, I agree, we are never completely under nor over a script." In the terminology I have developed here, Latour's suggestion implies that there is no clear distinction between *instructuring* and *transformative* agency. In practice, every statement has both these elements at the same time: "to organize is always to reorganize" (Latour 2008). When Hagerman created his bot, for example, he thought he was only reinscribing the status quo—but how wrong he was! Not only were the bot-statements highly transformative, they led to a chain of similarly transformative moments, such as the creation of Exclusion Compliance.

Users participate in the governance of Wikipedia by enacting instructuring and transformative agency; they act both *with* and *on* the project's organizing statements. It is in this sense that governing is not simply something that happens to people, cast down from above and exerting a paralyzing

---

46. While I won't go into detail here, I must stress that being above a script does not place oneself above *all scripts* or all framing statements. There is no absolute freedom or autonomy when acting on a script. Rather, the conditions for acting above a script are usually very specific and somewhat limited.

force; it is not that the written documents and the apparatus of material implements totally dominate the situation. Forceful statements flow through all three of these domains, zig-zagging between them and transforming along the way. Authority, as governing agency, is enacted within the project by embracing its framing statements. In other moments, users are compelled to act on these very framing statements: revising them; elevating one over another; expanding or reducing their domain; or continuing to affirm the validity of a statement's current existence. Agency is expressed in relation to and not in spite of the flow of statements. In light of this transformability of a formation, the project's most forceful statements—the ones that are most durable and most distributed—must be considered as such not because these statements have some miraculous autonomous force, but because they are continually enacted time and time again: "essence is the consequence, not the cause, of duration" (Latour 2008).

Governance *Ex Corpore*

The point of revisiting Weber and translating his account of the bureau, admittedly in quite a liberal fashion, is not at all to conclude that Wikipedia is in fact a bureaucracy after all or to side with others who have made this claim. Imposing this historical form would be squaring a circle. Rather, it is to provide a basis from which to pose the question once more of organization and of governance in the face of ad-hocracy and, by extension, openness. It is to think organizationally in regard to open projects. For while ad-hocracy contains the suffix "cracy," denoting government or rule (and derived from the Greek *kratia*, meaning "power" or "rule"), the concept of ad-hocracy actually says little of these things. In Toffler's original account, ad-hocracy emerges instead as the liberation from these things, as a theory of nonorganization. Ad-hocracy is less a theory of governance "for the purpose" and more a celebration of the *mere fact of changing purposes.*

Weber's triad of documents, the apparatus, and the body of officials serve as a convenient starting point for beginning to locate the most forceful statements—the ones that do the work of framing—in an organization. But whatever the ontological distinctions between this triad are, they are secondary to the task of tracing the forceful statements that flow between these different domains. It is the precise nature of these flows, their differing materialities and levels of performativity, and their distinctive organizing and sorting capacities, that constitute the contours of governance. Paying attention to the governing capacity of these statements doesn't allow one to locate in Wikipedia a new organizational archetype; there is no generalizable Wikiocracy.

Rather, it is the singularity of different organizational forms that such an approach accentuates. For example, while the organizational form of Wikipedia undoubtedly shares consistencies and regularities with FLOSS and other open projects, its precise form cannot be understood without considering the project's unique encyclopedic hinterlands; the specific take on truth and neutrality inscribed upon it by its cofounders; the many and varied operations of its bots; and the functionality of the MediaWiki software platform, among many other things.

To direct thought away from ad-hocracy and back to the *how* of organization and governance, to its specificity, I draw inspiration from a competing Latin notion, *ex corpore*. This expression commonly translates to "from the body," but can also mean "out of; by reason of; according to; as a result of" and "material; object; frame(work)." So, quite fittingly, *ex corpore* can also translate to "as a result of, or according to, the frame." *Ex corpore* points directly to the governing force of the frame, which is always materially inscribed in the order of things. To be clear, *ex corpore* is not in itself a mode of governance, not a "cracy"; it is a technique of description. *Ex corpore* is the practice of making visible the framing statements within a formation, the most forceful statements that constitute the governance of any organization. *Ex corpore* is the name I give to the method of positive description (of force), begun in the previous chapter and extended into the ones to follow; it is the method that makes it possible to rediscover the organization of openness. This practice can begin with a consideration of the three components of governance identified by Weber, by tracing their histories, their specific interrelations, and the unique dynamics that each singularity produces, but as noted this is only a convenient starting point to be translated and adapted however necessary (some organizations, for example, might have no written documents).

Ex corpore also has a specific meaning in the history of Catholicism regarding the veneration of sacred relics that is of relevance. In this use, *ex corpore* literally refers to a relic from the body, a piece of the body of a dead saint or martyr to be honored and worshipped. It is more than serendipitous that such an understanding, with its connotations of worship and the sacred, resonates with the treatment of written documents that govern the bureau, which are "preserved in their original or draught form" (Weber 1958, 197). *Ex corpore* therefore also points to the historicity of forceful statements, their continuity over time, their peculiar genealogical trajectories, but also to the fact that such a continuity is made possible through a direct and continuous relation with the present. Forceful statements remain so because they are honored time and time again, from present to present; it is not their history that makes them forceful, but their continuing force that makes them historical.

4

Organizational Exit and the Regime of Computation

> The core freedom in free software is precisely and explicitly the right to fork.
> STEVEN WEBER (2004, 159)

In chapter 1, I described the re-emergence of the political concept of openness. The aim of that chapter was to establish the force of openness as a new organizing mechanism, to show how it functions, what it is allied with, and finally to outline its limitations. I also showed that the open is remarkably scalable and can be applied more or less to any form of organization, from nation state to individual project. Indeed, openness has deep affinities with systems-based and cybernetic paradigms, and can therefore be applied to anything conceived in those terms. The open, however, with its allied notions of participation, collaboration and ad-hocracy, is not just an (ambiguous) mode of organizing open projects like Wikipedia. Rather, openness brings with it ideas about what constitutes the best way of organizing; it is an abstract theory of what constitutes the "good" of organization. We have already seen that collaboration was thought to be superior to older forms of working together because of its nonhierarchical nature, and the fact that it was not ordered by market price mechanisms or managerial commands. And we saw how ad-hocracy was celebrated over bureaucracy because of its ephemerality and the "associative" subjectivity it enabled. Focusing on this aspect of openness brings its associations with software and network cultures more closely into conversation with the long history of political thought dedicated to the question of just governance.

In this chapter I engage with the tradition of just governance in a more direct manner through an analysis of "forking" in open projects. In particular, I consider forking in relation to theories that deal with the question of leaving an organization or body politic no longer considered just. Previous iterations of this leave-oriented politics include writings about exit (Hirschman 1970), exodus (Virno 1996; Walzer 1985), escape (Papadopoulos, Stephenson, and Tsianos 2008), or even revolution. I show how forking differs from its historical counterparts and explore the various functions it serves within the political discourses of open projects; in particular, this takes place through a dialogue with the writings of Albert Hirschman. While writings about forking are highly idealistic and full of extraordinary claims, I do not dismiss them outright. Rather, it seems more important to attend to what makes them possible. The second half of the chapter therefore focuses on the relation between these writings and the realities they describe through a consideration of the Spanish fork of Wikipedia in 2002.

Legitimizing Organization

Every theory and practice of governance has its legitimating mechanisms. The question of governance is considered here parallel to that of the deliberate organization of a particular type of system. Considered as a system, it is possible to distinguish between three moments of governance, all of which pose the question (or problem) of legitimacy in distinct, if related, ways: the constitution of the system itself; the type of system; and the failure of the system. Why should there be governance rather than none?[1] What type is best? And what to do if governance turns bad? Tied up with these questions is of course an imaginary, liberal, sovereign subject in control of their destiny and a theory of power as something that rational individuals possess, control, and perhaps choose to delegate. Power is more or less equated with sovereignty.

In the writings of Hobbes, for example, the state itself—that is, the very constitution of a system of formal governance—is justified first and foremost by invoking a hostile, unforgiving, and violent state of "nature," where life is famously depicted as "solitary, poore, nasty, brutish and short" (1985, 185). Individuals agree to hand over their sovereignty, their natural freedom, in exchange for the security, peace, and order provided by a sovereign power. The state of nature is undesirable, but also individuals *agree* to leave it by way of the social contract.[2] Once this initial justification has been made, Hobbes

---

1. Indeed, this is exactly how Robert Nozick opens his libertarian treatise on governance: "The fundamental question of political philosophy, one that precedes questions about how the state should be organized, is whether there should be any state at all" (1974, 4).
2. The problem of the constitution of the state, and the invocation of a social contract as a response, was a major preoccupation of seventeenth- and eighteenth-century political thought. Rousseau similarly makes use of the social contract, although for different ends. As is well known, Rousseau's state of nature differs significantly from Hobbes's. There is nothing especially vicious or horrible about Rousseau's state of nature and he therefore justifies governance on different grounds. For Rousseau, an individual leaves the state of nature via the social

FIGURE 9. Top section of Bosse's frontispiece to Hobbes's Leviathan. Source: http://commons.wikimedia.org/wiki/File:Leviathan_gr.jpg (accessed November 3, 2013).

offers a series of justifications regarding the *type* of state, some of which can be quickly grasped by considering Abraham Bosse's famous frontispiece, which was drawn in close collaboration with Hobbes (fig. 9). The towering figure of the sovereign is legitimated through its representational character; it is comprised of the will of its citizenry. More specifically, the sovereign is comprised of its citizenry, dependent on it, but *more than* this mass of bodies or unruly multitudes. Its head and hands are distinct from the mass, suggesting a degree of autonomy, perhaps just enough to rule but not enough to rule without consent. The Leviathan's reach extends across the whole territory, as does its gaze. The sword in the right hand represents secular might, while the crosier in the right hand maintains an ambiguous but direct connection between the sovereign and the Divine. Imagined through this figure of the Leviathan, Hobbes's form of government mobilizes strength, divinity, omnipotence, and representativeness as sources of authority and legitimacy.

---

contract not for the protection of the sovereign who rules over them in a stern manner, but in order to attain a "virtuous" existence and to fully realize their capacity to become an "intelligent being and a man" (1986, 196). We see, then, the concern of the state and the same device mobilized, but in a different way and for different ends.

Interestingly, in the writings of Hobbes the question of what to do if government goes bad is not one a citizen can pose. There is no question of revolution (Hartman 1986, 494). The handing over of sovereignty must be absolute and unconditional. This unconditional submission was itself justified by the fact that Hobbes's political science claimed to provide the "Principles of Reason" that ensured a "constitution, excepting by external violence, everlasting" (378) and also, once more, by the absolute need to avoid the state of nature, or total war.

The question of what to do when a mode of organization is seen as unjust or in some way no longer suitable has, however, been considered often.[3] Most famously it emerges as the right of revolution or rebellion: if a leader fails to act in accordance with the will of the people, or alternatively if the leader violates the perceived "natural rights" of its subjects, they have the right and sometimes even duty to overthrow this leader and install a new one. As a response to such violent means of altering government, modern democracies have periodic elections as a way of giving citizens the chance to oust poor leaders without resorting to revolution.

At the level of organization, however, the strategies of citizens appear as just one of the possible responses to a particular type of organization. Different modes of organization respond to the question of (perceived) organizational failure, deterioration, or discontent in different ways. Perhaps no other writer has considered the general question of organizational failure as well as Albert Hirschman. In *Exit, Voice, and Loyalty*, Hirschman's focus is not so much on legitimizing certain forms of disruption but in understanding the different options available to members of an organization depending on the specificity of that organization. Through a consideration of private firms, two-party political systems, voluntary associations, families, and gangs, among other things, Hirschman argues that there are really only two options available to dissatisfied members of any organization. The first, which appears most readily in private organizations operating in markets, is "exit." If a consumer experiences a decline in the quality of an organization's output the consumer will go elsewhere. This idea of exit is equally extended to the actual employees and managers of the organization who might also become disillusioned (Hirschman 1970, 22–29). The second response, characteristic of organizations that are difficult to leave such as a nation or church, is "voice." Hirschman writes:

---

3. In response to Hobbes, see in particular John Locke's "Of Tyranny" and "Of the Dissolution of Government" toward the end of *Second Treatise of Government* (Locke 1976).

> To resort to voice, rather than exit, is for the customer or member to make an attempt at changing the practices, policies, and outputs of the firm from which one buys or of the organization to which one belongs. Voice is hereby defined as any attempt at all to change, rather than to escape from, an objectionable state of affairs. (30)

Read at the level of organization, exit (as the departure from the system) and voice (as an attempt to alter the system from within) emerge as two different forms of feedback, which signify problems in a system. And although Hirschman does not write a moral defense of exit and voice, he is interested in how through an optimal mix of these mechanisms an organization might return to its desired, optimal functioning.[4]

In what follows I argue that writings on different types of forking seek to position it, with varying levels of awareness, within this history of "legitimate organization." In particular, forking is mobilized in ways that strongly parallel Hirschman's writings on the perceived decline of organizations. Although for Hirschman the question of legitimization remains peripheral and largely implicit, in writings about forking it takes center stage. Indeed, I hope to show that the process or mere threat of forking is used to make legitimate a dizzying array of factors.

Forking

Forking is mentioned regularly in writings about software, but sustained engagements are few and far between. Much of these writings are in "native" online spaces, such as wikis, discussion forums, and home pages, and written by programmers themselves. I have tried to combine these writings with a more recent body of "scholarly" writings, which are scattered across the fields of legal studies, political science, media studies, and cultural anthropology.

While processes that resemble forking can be observed in many forms of organization, the notion of forking in open projects means something quite specific and draws directly from software culture. The term originally referred to an operating system process where the output of the process is a functional duplication of the process itself, thereby creating two separate but virtually identical processes. The translation of this technical definition into software

---

4. In this sense, Hirschman must be distinguished from the related notion of exodus developed by Paolo Virno and Antonio Negri. These writers suggest a generalized departure from the conditions of capital, a "collective defection from the state bond, from certain forms of waged work, from consumerism" (Virno and Costa, n.d.), not in order to restore capitalism to its optimum function, of course, but to destroy it.

and other content projects as a whole generally extends only to open projects since—because forking involves extensive and direct duplication—anything under the regime of copyright cannot be forked. Indeed, from an economic perspective forking directly contravenes the law of scarcity and seemingly, therefore, a central component of value under capitalism. This also means that forking is generally not considered applicable to "material" things such as hardware and traditional institutions, which also satisfy the scarcity criterion.

Definitions of forking vary in detail and content, but generally contain a few basic consistencies. The legal scholar Andrew St. Laurent writes that "forking occurs whenever a software project splits. While the two versions may remain entirely or partially compatible for some period of time, inevitably the unique (and now distinct) histories of each one's development will push them apart" (2004, 171). According to the MeatballWiki: "a fork happens any time development proceeds along two or more different paths." The fork "does not start from scratch, but continues to build on the resources of the project available up until the time of the fork."[5] The entry titled "Forking OfOnlineCommunities" on the same wiki adds that it is possible to fork software, but the community around it cannot be forked, only split.[6] As well as splitting the community, open source guru Eric Raymond stresses the competitive dimension of forking: "The most important characteristic of a fork is that it spawns competing projects that cannot later exchange code, splitting the potential developer community."[7] For the software developer and author David Wheeler, it is the intentional character of the resulting competition that is key: "What is different about a fork is *intent*. In a fork, the person(s) creating the fork *intend* for the fork to *replace or compete with* the original project they are forking."[8] Software anthropologist Chris Kelty offers this broad definition: "Forking generally refers to the creation of new, modified source code from an original base of source code, resulting in two distinct programs with the same parent," but he also shares Wheeler's understanding of forking as involving intent and competition: a "genuine fork" involves

---

5. *MeatballWiki*, "RightToFork," accessed October 5, 2010, http://meatballwiki.org/wiki/RightToFork.

6. *MeatballWiki*, "ForkingOfOnlineCommunities," accessed October 5, 2010, http://meatballwiki.org/wiki/ForkingOfOnlineCommunities.

7. E. Raymond, "Homesteading the Noosphere (Promiscuous Theory, Puritan Practice)," 2000, http://catb.org/esr/writings/homesteading/homesteading/index.html.

8. D. A. Wheeler, "Why Open Source Software / Free Software (OSS/FS, FLOSS, or FOSS)? Look at the Numbers!" (2007), accessed October 5, 2010, http://www.dwheeler.com/oss_fs_why.html.

"two bodies of code that [do] the same thing, competing with each other to become the standard" (2008, 139).

One of the most detailed discussions of forking is found on LWN.net. A highly specific and technical definition is put forward by a discussant, "jd," who begins by distinguishing four related but different processes of software development:

> A continuation—a project that continues the work of another after it has been discontinued;
> A clean-room implementation—a project designed to replace or compete with another, but doesn't borrow its code in any way;
> A multi-baseline implementation—competing sets of code within the same team/project;
> An extension or derivative—a project that extends upon a body of code, while leaving the initial code untouched.[9] (jd 2008)

Jd then defines forking in relation to these other processes: "A fork is any line of development that is not any of the four categories above AND is divergent to the point where direct bi-directional cross-pollination of unaltered code is either not safe or not meaningful" (jd 2008). For jd, therefore, a fork does more than continue a dead project; it doesn't simply perform the same *function* as its "parent," but actually makes use of the *same code*; this use is not designed to be fed back into the project (it isn't an internal development method); it isn't just an extra bit of code attached to a larger project; and finally, a fork becomes something different, something no longer compatible with the "parent."

From these various definitions, several common qualities of this technical process emerge: forking involves the splitting of a project, the duplication and reuse of existing code, and the creation of a distinct entity that ends up competing with the original project. Forking also brings into question the intent of the "forker" and the status of the wider "developer community," who cannot be forked in the technical sense of duplication.[10]

It is this relation between the people in the project—the "developer community" in software projects, but more generally any member or contributor in any open project—and the political significance they and other observers confer upon this seemingly technical process of forking that is of interest

---

9. jd, "Ten Interesting Open Source Software Forks and Why They Happened (Pingdom)," 2008, http://lwn.net/Articles/298092/.

10. It is also worth mentioning that forking is increasingly used in a more general way, such as when Kelty describes the Open Source Initiative as a fork of the Free Software Movement (2008, 99).

here. Such a focus on the political significance of forking is not random or at odds with the literature, and instead reflects the increased understanding of forking by those who write about it as a political process, especially as the overarching logic of openness is translated into new domains. In particular, I want to focus on two qualities of forking that are constantly put forward, what I call the *constitutive* nature of forking and its perceived function as a *safety net*.

The constitutive nature of forking in open projects is expressed in different ways. For Joseph Reagle it emerges as the test of the openness of communities: "I argue that a test of an open community is if a constituency that is dissatisfied with results of such a discussion can fork (copy and relocate) the work elsewhere" (2008, 75). For the programmer and author of *Producing Open Source Software*, Karl Fogel, forking is an "indispensable ingredient that binds developers together on a free software project" (2005, 88). Christian Siefkes, whose monograph *From Exchange to Contributions* attempts to show how the principles of peer production can be applied beyond informational goods, similarly writes: "If people want to leave a peer project or local association, they are free to go. This freedom to leave or to 'fork' is an essential aspect of free cooperation" (2008, 121). There are also versions of this constitutive statement of forking that are expressed more explicitly in the language of politics. The MeatballWiki, for example, defines forking in the tradition of rights: "The right to fork is inherent in the fundamental software freedoms common to all open source software."[11] Political scientist Steven Weber, whose book *The Success of Open Source* focuses explicitly on the governance of these projects, places forking in the related tradition of "freedom": "The core freedom in free software is precisely and explicitly the right to fork" (2004, 159). Forking, then, is crucial to the identity of open projects, so much so that a project that cannot be forked is not considered open. And although all these writers ascribe fundamental and hence constitutive importance to forking, for Weber forking sits at the very "core." It is more important, more definitive, than any other quality or "freedom."

The second key quality of forking has to do with its function within open projects. I call this function, which has less to do with the actual process of forking and more to do with the effect of invoking the possibility of forking, the "safety net" function. Forking as safety net refers to the "final option" or "last resort" of the disgruntled, the marginalized, or the otherwise unhappy in open projects. Participants who do not like the way a project is governed (the direction it is taking, the decisions of the core group, the way the project

---

11. *MeatballWiki*, "RightToFork."

has developed over time, or the members involved) and are unable to convince key decision-makers of the need for change can as a last resort leave the project taking the source code with them and realize their competing vision. It might seem that the safety net function is no different from forking itself, but this is not the case. The safety net is more than a process; it is a mechanism of legitimization. The main function of the "fork as safety net" is to guarantee the legitimacy of the governing mechanisms of all open projects. The safety net function seemingly ensures that there are no great injustices, that everyone involved in a project must, in the last instance, support how things are done. Even if an actual fork never takes place (an event that would both challenge the legitimacy of governance and simultaneously demonstrate or ensure this legitimacy on a higher level), the mere threat of forking is enough. Indeed, fork as safety net has little to do with actual forking. As a mechanism of legitimization, its function is directed inward and its effect is measured through the effect it has on the governance and general organization of the project. Forking as both *constitutive* and *safety net* sits interestingly alongside Hirschman's notions of exit and voice and I will explore this relation in a moment. First, however, I want to delve a little deeper into this safety net function. If fork as safety net has less to do with the process of leaving and is instead first and foremost a mechanism of legitimization, the questions that need to be posed are: What does the safety net function do? What does it make legitimate? Or what kind of just organization does it ensure, if not bring entirely into existence?

Perhaps the most general iteration of the safety net function, and one that comes closest to gesturing to the outside of a project, is provided by the legal scholar James Boyle, who writes, "In the last resort, when they disagree with decisions that are taken, there is always the possibility of 'forking the code,' introducing a change to the software that not everyone agrees with, and then letting free choice and market selection converge on the preferred iteration" (2008, 190). At the most basic level, the safety net allows for the exit of participants. This freedom of association is perhaps the most fundamental. However, it also guarantees a kind of agency—a capacity to alter the code—that is somehow removed from the formal mechanisms of authority. Finally, once more it is the market, upon which these two competing products are cast, which forms another crucial, if deferred, ground for legitimization. If all else fails the market will decide who and what is right, but more than this, it is the mere fact of the market itself, the potential to compete freely, that is the basis of legitimacy. So even in this very general statement of the safety net function there are actually three components at work: the right to leave; the capacity to modify code; and the potential to defer to the market, that is, to a different and ultimately infallible process of determining what is true and just, which combine to ensure that, overall, the project is governed in a legitimate manner.

Terry Hancock, who writes for *Free Software Magazine*, situates this most general function of forking squarely within the history of mechanisms for political exit. Hancock concludes his commentary on the fork of *OpenOffice.org* by situating forking within the history of rights, freedom and "tools of revolution." He writes:

> The situation [of forking] reminds me of some other controversial freedoms in the real world: the freedom of expression, freedom of assembly, and the right to bear arms guaranteed in the US Constitution's Bill of Rights, for example. Certainly these are not universally understood or agreed upon, but it is interesting to think about why they are there.
>
> What they are, really, are tools of revolution, put in the public's hands as a guarantee against the worst excesses of government. 'After all,' we may imagine, 'if it gets too bad, we can just revolt.' Invariably, cooler heads prevail, but the virtual threat of such an action does keep the government from getting too far outside of its democratic mandate.
>
> Likewise, the freedom to fork a free software project is also a 'tool of revolution' intended to safeguard the real freedoms in free software. Every once in awhile someone comes along who tries to abuse the illusory power over the community, and a little revolution has to occur (fortunately, in the case of software, this is almost always a bloodless revolution!). As such, it's a precious right that we should not give up.[12]

In addition to this general function of the safety net, there arise a series of more specific functions in relation to power and governance. Or more accurately, this general function is endowed with a series of more specific effects. For example, Karl Fogel writes:

> Forks, or rather the potential for forks, are the reason there are no true dictators in free software projects. . . . Imagine a king whose subjects could copy his entire kingdom at any time and move to the copy to rule as they see fit. Would not such a king govern very differently from one whose subjects were bound to stay under his rule no matter what he did? . . . This is why even projects that are not formally organized as democracies are, in practice, democracies when

---

12. T. Hancock, "OpenOffice.org Is Dead, Long Live LibreOffice—or, The Freedom to Fork" (2010), *Free Software Magazine*, http://www.freesoftwaremagazine.com/columns/openoffice_org_dead_long_live_libreoffice.

it comes to important decisions. Replicability implies forkability; forkability implies consensus. (2005, 88)

The "potential for forks" is here situated in relation to dictatorship. It stands in as a defense against tyranny and a guarantee of democracy, whatever the "formal organization" of the project may be. Christian Siefkes's definition of forking includes a similar democratic function. He writes, "People 'vote with their feet,' supporting the projects they like, and leaving or forking a project if they are unhappy about the way in which it evolves" (2008, 68).

But forking, at least for Fogel, ensures more than democracy, however conceived. "Consensus" emerges as a post- or more-than-democratic harmony. If for agonistic writers such as Chantal Mouffe (1993), the best democracy can hope for is the recognition that the actual *process of mediation* between opposing sets of desires, and therefore the creation of winners and losers, is itself seen as valid—that is, that all parties recognize the frame as legitimate—forking seems to ensure a more harmonious outcome. For Mouffe, a general condition of harmony is not achievable because the social is constituted by and continually produces possible antagonisms. Political antagonism can only ever be mediated—therefore becoming agonistic as opposed to antagonistic—and never resolved.[13] Put in the language of Lyotard, "damages" result from agonistic encounters and "wrongs" from antagonistic ones. For Fogel, the "replicability" of free software projects means that any "loser" of a formal process, regardless of whether or not the procedure is itself seen as fair, can leave to preside over his or her own kingdom. All kingdoms are therefore harmonious utopias where everyone is in agreement. At the extreme end of consensus, however, lies a world of a million kingdoms, each populated by a single (happy) king or queen. Agonism is replaced with radical separation.

---

13. On politics and consensus Mouffe writes, "Politics . . . has to do with conflicts and antagonisms. It requires an understanding that every consensus is, by necessity, based on acts of exclusion and that there can never be a fully inclusive 'rational' consensus" (1993, 141). To support this position Mouffe draws on Derrida's discussion of what she describes as the "constitutive outside": "The constitution of any identity, whether individual or collective is always based on excluding something and establishing a violent hierarchy between the resulting two poles—form/matter, essence/accident, black/white, man/women, and so on. This reveals that there is no identity that is self-present to itself and not constructed as difference, and that any social objectivity is constituted through acts of power. It means that any social objectivity is ultimately political and has to show traces of the exclusion which governs its constitution, what we can call its 'constitutive outside.' As a consequence, all systems of social relations imply to a certain extent relations of power, since the construction of a social identity is an act of power" (141).

Steven Weber, Terry Hancock, and Michel Bauwens all claim that forking effects a radical transfer of power from leaders to followers. This radical transfer is then used to reflect back on leadership, which becomes precarious and meritocratic. Weber writes,

> By creating the right to fork code, the open source process transfers a very important source of power from the leader to the followers. The privileges that come with leadership then depend on the continuous renewal of a contingent grant from the community. This comes as close to achieving practical meritocracy as is likely possible in a complex governance situation. (2004, 181)

Once again commenting on the fork of *OpenOffice.org*, Hancock writes,

> Apparently the management of the company has learned their strategy from proprietary software, where a political executive decision can kill a project, regardless of developer interest. But with free software, the power lies with the people who make it and use it, and the freedom to fork is the guarantee of that power. (2010)

Here the precise effect of the transfer of power is that managers cannot control the life and death of a project. It also serves to differentiate the managerial style of free software from commercial projects and their strict chains of command.

Bauwens's position is a little more elaborate. The safety net function is linked to what he calls the "de-monopolization of power" and "maximum freedom" to be found in his vision of a P2P civilization. However, his sentiment is very similar to Weber's "meritocracy" when he argues that power in P2P processes takes the form of "reputation that generates influence." This power is "given by the community" and "time-bound to the participation of the individual." Immediately after Bauwens writes: "In the case where monopolization should occur, participants simply leave or create a 'forking' [sic] of the project, a new path is formed to avoid the power grab" (2005a, 36). Forking therefore not only de-monopolizes power, but also acts as an antidote to its excesses.

This negative action has its positive equivalent in the realm of freedom, which also brings Bauwens closer to Fogel and Seifkes:

> P2P is predicated on the maximum freedom. The freedom to join and participate, to fully express oneself and one's potential, the freedom to change course at any point in time, the freedom to quit. Within the common projects, freedom is constrained through communal validation and consensus (i.e. the

freedom of others). But individuals can always leave, fork to a new project, create their own . . . Unlike in representative democracy, it is not a model based on a majority imposing its will on a minority. (Bauwens 2005a, 39)

For Bauwens there is an ease to forking: if unhappy, "participants simply leave or create" a fork because of their maximum freedom, and this can happen "at any point in time." Along with this power-reducing, freedom-enhancing safety net function, for Bauwens forking also plays a crucial role in distinguishing open projects (or in this case P2P) from the organizational form of representative democracies. What allows Bauwens to write that P2P "is not a model based on a majority imposing its will on a minority" is precisely the capacity to exit and fork.

Finally, Mark Elliott uses forking to refute Jaron Lanier's influential critique regarding the pitfalls of online collaboration. In his controversial article, "Digital Maoism," Lanier argues that projects such as Wikipedia are a hotbed for "foolish collectivism" (2006). In response, Elliott writes, "In wiki contributing contexts, instead of collectivism manifesting, individuals with differing perspectives tend to 'fork' projects, as the ease of creating new mass collaborations is relatively easy" (2007). Steven Weber makes similar reference to diversity when he writes: "Forking, like speciation, is an essential source of variation and ultimately of radical innovation" (2004, 169). Admittedly, this last gesture toward creativity and diversity is not directed inward; it does not legitimize individual projects. It does, however, privilege open projects on a more general level. It also reveals how versatile the safety net function of forking can be.

As both an actual process and a rhetorical device, forking sits interestingly in relation to Hirschman's notions of exit and voice. Forking emerges first as a kind of "exit," but with benefits. If a person perceives a "decline" in an organization (project), if she has a major disagreement, she simply duplicates the code and proceeds in a different direction. Following on from this, as a direct result, "voice" in open projects is thought to be especially powerful: If there is nothing keeping the participant (because she can take "everything" if she leaves), existing project leaders had better listen or else risk losing the contributor base. For Hirschman, different organizations respond differently to each mechanism and the task is to try to attain the right mix to ensure that short-term decline remains exactly that. The balance of exit and voice is also subject to change as organizations change over time and in response to previous arrangements of these very mechanisms. Compared to other forms of organization, in open projects the character of exit (lossless and ever-present) and voice (almighty) is quite unique.

As we have seen, however, the literature on forking is not focused on the balance of exit and voice, but on the particular dynamics produced by their unique relation to each other: on how the lossless, ever-present character of exit bears directly upon that of voice, and how this relation in turn produces the array of "legitimization effects" I have considered above. Stemming from the safety net option—from the right to leave, the capacity to modify the code, and the option to defer to the market—(the threat of) forking is a defense against tyranny and guarantor of democracy, it produces a form of consensus, transfers power from leaders to followers, achieves practical meritocracies, de-monopolizes power, ensures maximum freedom, and brings about diversity and radical innovation. As a unique mode of exit, I am tempted to conclude that forking is a truly wondrous thing, imbued with an astounding set of capacities. As a particular form of organization that responds to the moment of organizational decline via the threat or implementation of forking, it would seem that the governance of open projects is always and necessarily legitimate. And because forking is constitutive of openness, this *a priori* legitimacy extends to all things open.

Forking and Computationalism

What is it about forking that makes it so special? That makes exit so bearable and voice so powerful? I have already gestured to forking's "lossless" quality and to its existence in a realm of perceived abundance. And we saw that the power of voice derives directly from these conditions as well. In the second part of this chapter I look more closely at what makes such a perspective possible, in order not only to bring to light the unique conditions from which this form of political action emerges, but also to find out something more about the relation between descriptive statements about things (the discourse on forking), these things themselves (the process of forking), and the kind of translations and processes that are necessary for the two to reach alignment. The focus will not be on the performativity of written statements about forking, but on the conditions that need to be established in order for certain statements to acquire a truth-value and general acceptability.

The rising influence of computation across many areas of life is well documented. By computation, I don't simply mean computers or technology, but a particular flow of computational forces (of computational statements) that is manifested in technologies, practices, writings and politics. For writers like N. Katherine Hayles, the "regime of computation" provides an entire worldview, complete with its own metaphysics, and she sees this regime at work across a variety of milieux. She writes:

> The Regime of Computation . . . provides a narrative that accounts for the evolution of the universe, life, mind, and mind reflecting on mind by connecting these emergences with computational processes that operate in both human-created simulations and in the universe understood as software running on the "Universal Computer" we call reality. (2005, 17–19)

The primary expression of computation, something akin to its prototypical apparatus or diagram, is the Universal Turing Machine, "a device that can perform any computation that any other computer can do, including computing the algorithm that constitutes itself" (17–19). Hayles continues:

> The computational regime continues in the tradition of Turing's work by considering computation to be a process that starts with a parsimonious set of elements and a relatively small set of logical operations. Instantiated into some kind of platform, these components can be structured so as to build up increasing levels of complexity, eventually arriving at a level of complexity so deep, multilayered, and extensive as to simulate the most complex phenomena on earth, from turbulent flow and multiagent social systems to reasoning processes one might legitimately call thinking. (17–19)

Hayles also points to a productive ambiguity within this regime that I want to take up later: namely, how computation oscillates between "metaphor and means" (17–19), that is, between a mode of understanding and a statement about how the world actually is.

Resonating with the work of Hayles is David Golumbia's *The Cultural Logic of Computation*. Golumbia is similarly interested in the widespread "belief in the power of computing" which he calls "computationalism" (2009, 2). Rather than stressing the novelties of computationalism, Golumbia sees it as the latest iteration of the longstanding belief in "something like *rational calculation*" or a "rationalist theory of mind" (1). This historical focus and link to rationalism proves valuable in his discussion of "computational politics" because it allows him to identify consistencies between popular refrains in computational discourse and longstanding themes of liberal, individualist politics, such as "computers empower users" (181). I also want to explore something like "a regime of computation" or "computationalism" at work in understandings about forking, but take a slightly different approach to these authors. By keeping my analysis specific, I want to generate equally specific and hence novel accounts of the contours of "computationalism." Furthermore, while I share Golumbia's deep suspicion of the powers of computation I want to account for how such a belief comes to be accepted and materialized in everyday practices and also try to establish the real limits of its operation. It is a question of how computationalist realities become real, how constant feedback loops between things, words, bodies, and processes work to (re)create and affirm computational realities.

If questioned on whether or not forking is actually possible, any FLOSS programmer or peer production activist could quickly rattle off a large list of examples from both software and post-software (content) projects, beginning perhaps with the famous fork of the GNU C Compiler (GCC), which produced the more successful Experimental GNU Compiler System (EGCS).[14] Operating systems, productivity suites, content management systems, compilers, web browsers, content projects (such as encyclopedias), and even entire movements (FSM/OSI) feature on the list of things successfully forked. Any reorientation of forking and any interrogation of its remarkable legitimizing capacities must account for these empirical realities.

I noted earlier that the term *forking* stems from a technical process in an operating system, whereby the output of the process is a functional duplicate of the original process. A series of code statements are duplicated and obtain a functional autonomy from the original statements. With this initial definition there is no "project" per se, only a string of code running on a computer. When the notion of forking was translated to include entire software projects, the focus nonetheless remained largely on the (source) code. This is not surprising, considering that the output of such projects was predominantly software and a large part of participating in such projects takes place via contributing lines of code. As I illustrated with Stallman's struggles in chapter 1, source code is bound up with the limitations and capacities of programmers. Code becomes a *source* of agency and politics more generally. While not all open projects contain *source code*, they nonetheless all have a *source*. Indeed, as a concept forking is completely dependent on the idea of a source and every fork is simultaneously an expression of that source.

Recently, however, Wendy Chun has pointed to the limitations of singling out source code as an object of power and knowledge. Although Chun is sympathetic to the turn to software within studies of new media, she nonetheless argues "that positing software as the essence . . . [or] source relies on

---

14. This fork originated out of a disagreement between GCC's official project maintainers and a group of highly active developers from Cygnus Software. These developers created the EGCS fork in order to increase the speed of development and incorporate new patches into the code more regularly. The fork was so successful that it became more popular than the original compiler. Recognizing this, the original GCC maintainers reincorporated the improvements of EGCS, thereby unifying the compiler once more under a single name (see Fogel 2005, 225–26). (Interesting, this ability to reincorporate the code base of a fork goes against many of the definitions provided as to what constitutes a fork.)

a profound logic of 'sourcery'—a fetishism that obfuscates the vicissitudes of execution" (2008, 300). Echoing claims made by Golumbia, she continues:

> this sourcery is the obverse rather than the opposite of the other dominant trend in new media studies: the valorization of the user as agent. These sourceries create a causal relationship among one's actions, one's code, and one's interface. The relationship among code and interface, action and result, however, is always contingent and always to some extent imagined. The reduction of computer to source code, combined with the belief that users run our computers, makes us vulnerable to fantastic tales of the power of computing. (300)

Chun provides a complex engagement with fetishism, but I will focus on two related elements that are particularly relevant.[15] First, fetishes establish a "'unified causal field' that encompasses both personal actions and physical events." In this case, they establish a causal relationship between one's action and the execution of code. Second, the establishment of this causal field "relies on distorting real social relations into material givens." Code becomes "a magical entity—as a source of causality—when in truth the power lies elsewhere, most importantly in social and machinic relations" (300). In sum, Chun's "sourcery" involves singling out one entity as *the* source of knowledge and power. By overlooking "the vicissitudes of execution" and the ensemble of "social and machinic relations" required for such execution, a causal relation between user and source can be established. This in turn secures the agential capacities of users, who attain power and enlightenment through knowing and mastering the source.

While Chun's critique centers on the gap between source code and execution, a similar "logic of sourcery" is at work in writings about forking: these, too, single out a project source in the face of distributed and uncertain political realities and also overlook the gap between a desire to fork and its realization or "execution." However, I want to extend and perhaps complicate this logic of sourcery by suggesting that the fetish is not limited to the domain of users. Instead, sourcery is inscribed in the very materialities of computation; it is the grand accomplishment of a whole series of conspirators. Seeing how this is achieved will tell us something about the means and metaphors of forking—and computation more generally.

---

15. It is at least worth mentioning, however, that Chun does not see knowledge or truth as the answer to this software fetish—indeed, the fetish is founded on claims to enlightenment—and she fully recognizes that the fetish is known and realized by its participants, but practiced anyway. Her response is therefore not to escape the fetish, but to "make our interfaces more productively spectral" (2008, 300).

Even in its original technical definition, forking is never one-to-one. The relation between the two processes is never one of perfect equivalence. Although the processes appear "functionally identical," in small and seemingly insignificant ways, they are different. Rather, every act of forking is a translation as well as a continuity. Matthew Kirschenbaum gets at this in his methodo-ontological distinction between "forensic" and "formal" materiality. For Kirschenbaum, "forensic materiality rests upon the principle of individualization (basic to modern forensic science and criminalistics), the idea that no two things in the physical world are ever exactly alike." Forensic materiality stresses the irreducibility of things to one another. He continues: "If we are able to look closely enough, in conjunction with appropriate instrumentation, we will see that this extends even to the micron-sized residue of digital inscription, where individual bit representations deposit discreet [sic] legible trails that can be seen with the aid of a technique known as magnetic force microscopy" (2008, 10).

Regarding the technical process of code forking, forensic differences exist at this level of digital inscription, as well as in the position of the two entities on the storage device, something that is acknowledged in the process itself via the creation of a unique "address space" for the new (forked) process. However, the forensic materiality of code is not just about identifying trace differences; it invites us to attend to all forms of difference, from all but undetectable variations in the process of magnetic binary inscription to different labor practices, temporalities, methods of production, storage, the different kinds of technological waste that result from these practices, and so on, that could be properly understood as ecological or network transformation. Paraphrasing an observation made by Latour, the forensic method never sees information, only transformation (2002).

"Formal materiality," writes Kirschenbaum, "is perhaps the more difficult term, as its self-contradictory appellation might suggest." Formal materiality refers to the consistencies that exist or perhaps "persist" across forensic difference: "Whereas forensic materiality rests upon the potential for individualization inherent in matter, a digital environment is an abstract projection supported and sustained by its capacity to propagate the illusion (or call it a working model) of *immaterial* behavior: identification without ambiguity, transmission without loss, repetition without originality." Formal materiality, we might say, is concerned with information and habitually backgrounds its transformations. Importantly, differences at the forensic level constantly work against the realization of formal consistencies. Formal materiality is never a given; it has to be achieved. Kirschenbaum notes, for example, how all "forms of modern digital technology incorporate hyper-redundant error-checking routines that serve to sustain the illusion of immateriality by detecting error and correcting it, reviving the quality of the signal" (2008, 11).

Kirschenbaum is therefore correct to describe formal materiality as a "working model." In one sense, "working model" implies a certain practicality and perhaps fallibility—the formal perspective is used because it works, not because it is true. In this regard, it is similar to Chun's "sourcery." In another and more important sense formal materiality is a working model because it is constantly *working*: the formal model has to be constantly maintained. The illusion of immateriality—of the creation of perfect equivalence or "lossless forking"—is underpinned by a cast of actors and processes, such as "hyper-redundant error-checking routines," which form the *forensic reality* of abstract formalisms. The ability to single out "source" at the expense of the vicissitudes of execution, the various "levels" of codic translation (e.g., compiler, assembler, binary) as well as the network of "social and machine relations" that need to be in place in order to maintain the fiction of causality between user, source, and execution, is itself an effect of the successful alignment of all elements in this "working model." The working model of formal reality, with its "ensemble of conspirators," makes something like the sourcery identified by Chun possible.

As a concept that emerges from the practice of programmers, in both its strictly technical and extended sense, forking is also underpinned by a formal understanding of digital media; forking is about duplication and the creation of equivalences. By extension, political investment in forking is also predicated on the ability to maintain this "illusion" of equivalence in the face of differences at the forensic level. It is clear though that as forking is attached to more-than-technical processes, the gap between the formal and the forensic, as well as what is at stake in this gap, is significantly altered. Attending to an actual example of a fork will help tease out these differences, their significance, the nature of forensic resistances and how (if at all) they are overcome.

The Enciclopedia Libre Universal en Español

At this point I want to introduce an event that will take center stage in the next chapter: the 2002 Spanish fork of Wikipedia. My focus in chapter 5 will be on the details leading up to the fork. For now, though, I want to skip those details and jump right to their aftermath, when a prominent figure of the Spanish Wikipedia community, Edgar Enyedy, was faced with the task of actually creating a fork, now known as the Enciclopedia Libre Universal en Español (EL).

Enyedy had emerged from heated debates with Wikipedia's founders regarding the current and future operations of the project with the Spanish community on side and they encouraged him to take action. Because all Wikipedia (source) content was licensed under the GNU/FDL at the time, it could be copied and reproduced elsewhere—it could be forked. Enyedy reflects on the process:

> Setting up the new encyclopaedia wasn't an easy job. I began by configuring a spare PC as an Apache server and started working on the software. The Perl scripts ran OK and the wiki could be reached through a proxy server from other computers on the net. "Well," I thought, "it runs." It took me a week to get it going, but this seemed a very small amount of time when compared to the dozens of hours I spent arguing about the project with Wales and the community. (Enyedy and Tkacz 2011, 116)

Getting the UseModWiki software installed and running was a major step, but there were several others. Enyedy also recounts, for example, his attempts to export the source content:

> At that time, to set up a wiki and to export the .tar database from Wikipedia was almost impossible. The GNU/FDL license granted it could be done, made it legally possible. But no way! The Wikipedia page on Sourceforge had instructions that read like hieroglyphics. And once again due to "technical" reasons (that none of us believed), the downloadable database was never updated. (116)

Exporting the database via the Sourceforge instructions eventually proved too difficult, especially since it wasn't updated (meaning it didn't contain the most recent version of the encyclopedia). Eventually Enyedy and the other "forkers" reverted to less sophisticated methods: "When the server was up and running, and as the GNU/FDL permitted, we began copying our articles from Wikipedia. It wasn't an automated process, no bots or anything, just us bringing the articles across one by one from Wikipedia's server to ours" (116).

Enyedy also had to find a home for the new project:

> The first thing I thought about was looking for a hosting company and registering a domain. I was also thinking about how we could make this component effectively community-owned. I had the idea, for example, that we could change the domain registrar each year so there was not a single continuing owner. There were few hosting companies with the characteristics I was looking for. Remember, at that time, to work on the server side was not as usual as it is today . . . we ended up getting hosting from the University of Seville. (116)

Part of the debate that led to the fork was about Wikipedia's hosting with a commercial entity, Wale's small company Bomis, and its .com domain name, so it's not surprising that these were points of difference. At the same time, these differences had to be "working" alternatives or contain enough functional similarity to be accepted by the community.

Enyedy's actions can be understood as a kind of "making equivalent," as the establishment of a new "working model" that makes possible the fork itself. These same actions are what make possible political writings about forking, underpinned as they are by a "formal," computational orientation. In order for the fork to become real, Enyedy had to install the wiki software, set up a server, transfer the source content, find a suitable host and new domain, and, of course, secure a body of contributors. During this process, Enyedy faced different types of resistance: hieroglyphic instructions; outdated databases; server set-up and software installation issues; content transfer difficulties (which involved the tedious manual task of copy-pasting content); finding suitable hosting; solving domain dilemmas and so on, which all worked against realizing the EL. Enyedy was able to overcome these (forensic) resistances, or find working alternatives for those that proved too strong, such as the task of exporting the .tar database. During the process, his programming skills were essential as was the help of the community in manually exporting articles. This help was itself only available because Enyedy was an unofficial leader of the community and had emerged successful from the earlier conflict. We can think of these elements as his allies in the process of securing the working model of formal equivalence.

The EL fork brings together the two main strands of this chapter: writings about the mode of organizational exit unique to open projects and questions about the reality of computationalism. In particular, the EL fork shows the relation between these writings and their realities—how computationalism attains a level of reality that in turn permits a corresponding discourse. Focusing on this historical event shows that the sourcery described by Chun, for example, depends on a *corresponding reality*. If focusing on "source" at the expense of more distributed relations of force is a fetish, it is a predesigned and more-than-human fetish—a relational effect of invisible or habitually overlooked actors working in unison. In the language of Kirschenbaum, there are always forensic realities to formal sourceries.

Juxtaposing writings about forking with the EL fork therefore offers insights regarding the "crucial ambiguity" within the regime of computation introduced by Hayles: whether it should be understood as a "pervasive metaphor" or "whether it has ontological status." Hayles wisely avoids trying to settle this ambiguity and prefers to see the two as "inextricably entwined" and as "a generative cultural dynamic" (2005, 20). This last phrase captures the complicated existences and transformations of forking, as thing, metaphor, thing metaphor. . . . Forking is always already more than a mode of understanding and always "working" to be more real, to be real-ized. This process is at work from the initial translation of forking from technical process to software projects, and then from software projects to content projects. Each movement has both metaphoric and ontological dimensions at the same time, to the point where the distinction is no longer relevant, except perhaps as a way of tracing the flow of statements about forking (formal, computational statements) as they become more durable, loaded with forensic materialities.

To emphasize the processual dimension of forking, the establishment of a working model of formal equivalence, is also to point to its contingent nature. This contingent nature in turn bears directly on writings about forking as a mechanism of legitimization. If the success of forking is based on the creation of a working model, which is subject to a host of resistances beyond (as well as within) the internal dynamics of a (preforked) project, then its function as mechanism of legitimization is itself contingent upon realities that cannot be known in advance. This is not to suggest that forking is not possible, or that it hasn't ever been achieved, rather that it is never guaranteed and neither therefore is the function of forking as a mechanism of legitimization.

I want to end this chapter by considering the contingency of forking in relation to questions of scale. I think this relation bears especially on writings that see open projects as potentially replacing other forms of large-scale organization. The Spanish fork of Wikipedia (the EL) was generally considered a success.[16] Enyedy managed to create a working alternative and, together with the rest of the (former) Spanish Wikipedia community, continued to pursue their encyclopedic vision. At that time, the Spanish Wikipedia was a very small project. There were a little over one thousand articles to transfer and maintain. According to Enyedy, there were only twenty to twenty-five regular contributors and perhaps thirty-odd more who contributed infrequently (Enyedy and Tkacz 2011, 111). While Enyedy and the others faced difficulties in setting up the fork, they were able to overcome or route around these without too much trouble. Enyedy had enough allies to overcome any problems. In this sense, although the fork was not a certainty the chances of not securing formal equivalence were relatively small.

16. I am referring here to a success in *establishing* the fork. The EL has not stood the test of time and is no longer in genuine competition with the Spanish Wikipedia.

When Realities Scale

As open projects like Wikipedia persist over time and space, they garner new participants, add content (in this case entries, discussion pages, and so on), develop and argue over rules and policies, secure regular funders, become embroiled in media scandals, set up national chapters, celebrate milestones, and generally extend outwards and thereby become more real. Their forensic reality is amplified; their boundaries grow, shift, and are difficult to locate. When projects scale, so too does the difficulty in achieving a working model of formal equivalence. Difference is everywhere. The trick of sourcery is revealed. As of February 2012, for example, the English Wikipedia is the largest of all Wikipedias. It has over 3.86 million articles and over 26 million pages in total, almost 150,000 registered users considered active, and 692 active bots.[17] In total, the project requires hundreds of servers, which are spread across Asia, Europe, and North America.[18] Alexa currently ranks Wikipedia as the sixth most popular site in the world and it regularly tops most search engine results.[19] The foundation that oversees Wikipedia employs over sixty people and has an annual operating budget over US$20 million—a figure that has steadily increased each year. What kind of working model is required to fork something of such scale?

Writings about forking tell us how people understand their practices and justify their actions. They tell us something about underlying dreams and desires, about visions of what might be and what has just arrived. In order to draw out the uniqueness of these writings, I have placed them in conversation with long-standing questions of governance and organization. Forking emerged as a kind of lossless exit whose unique nature formed the basis for a wide array of utopian political situations, from transferals of power to the community, to ensuring decisionmaking is always based on consensus. Rather than dismiss these writings offhand, I tried to understand the conditions that make them possible. This led to more complex questions about the relation between political writings informed by the "regime of computation" and its corresponding reality. Through a consideration of the EL I tried to show the translation of a computational logic (of forking) into new situations, a process I described as contingent and whose success I based on the

---

17. *Wikipedia*, s.v. "Special Page: Statistics," accessed February 1, 2012, http://en.wikipedia.org/wiki/Special:Statistics.

18. *Wikipedia*, s.v. "Wikipedia Contributors," accessed February 1, 2012, http://en.wikipedia.org/wiki/Wikipedia_servers#Software_and_hardware.

19. "Top Sites," accessed February 1, 2012, http://www.alexa.com/topsites.

possibility of installing a "working model" of formal equivalence in the face of (forensic) resistance. Pointing to the contingency of this process and the work required in perpetuating formal realities in turn cast doubt on political writings about forking. This approach allowed me to affirm the reality of forking (its past successes) and its political discourse, while also pointing to its real limitations—limitations that increase significantly as a project scales. Even though writings on forking have thus far depicted it in a rather straightforward and settled manner, I hope to have shown that much like Marx's commodities, forking too abounds in "metaphysical subtleties" and I continue to explore these in the next chapter.

5

Controversy in Action

February 7, 2002, was a memorable day for the Spanish Wikipedia. A prominent community member, Edgar Enyedy, posted a brief message to the international Wikipedia discussion list, noting that the Spanish Wikipedia had reached 1,000 article pages.[1] The achievement was met with congratulations from the English Wikipedia cofounder Larry Sanger, general back-patting among the Spanish community members, and invitations to share insights with other language Wikipedias on how they had achieved such rapid growth. The Spanish Wikipedia, it seemed, was a shining example among the host of new Wikipedias that had sprung up shortly after their English counterpart.[2]

Less than a week later another exchange began between Enyedy and Sanger, this time of a very different tone. In part of a longer post announcing the end of his paid employment by Wales's company Bomis, Sanger mentioned in passing that "Bomis might well start selling ads on Wikipedia sometime within the next few months."[3] Sanger's hope was that selling ads would generate enough revenue for him to return to his paid editorial position at Bomis. To this possibility of selling ads on Wikipedia, Enyedy replied:

> I've read the above and I'm still astonished. Nobody is going to make even
> a simple buck placing ads on my work, which is clearly intended for com-

---

1. E. Enyedy, "Spanish Wikipedia Meets the Challenge" (2002), accessed March 3, 2011, http://osdir.com/ml/science.linguistics.wikipedia.international/2002-02/msg00018.html.

2. The English Wikipedia was launched on January 15, 2001, and the Spanish version four months later, on May 1, 2001; *Wikipedia*, "Spanish Wikipedia," accessed February 13, 2011, http://en.wikipedia.org/w/index.php?title=Spanish_Wikipedia&oldid=409905416.

3. L. Sanger, "Announcement about My Involvement in Wikipedia" (2002), http://osdir.com/ml/science.linguistics.wikipedia.international/2002-02/msg00037.html.

munity, moreover, I release my work in terms of free, both word senses, I and [*sic*] want to remain that way. Nobody is going to use my efforts to pay wages and or maintain servers.

And I'm not the only one who feels this way.

I've left the project.

You can see the Spanish Wikipedia development in the last two days and then you may think it over.

Good luck with your wikiPAIDia

Edgar Enyedy

Spanish Wikipedia.[4]

On February 26, two weeks after this second exchange, the majority of the Spanish contingent abandoned the Spanish Wikipedia.

In the previous chapter, I considered the tail end of this event—the actual task of setting up the new encyclopedia (the EL). In this chapter, my focus is on the two-week period between Sanger's initial remarks about advertising and the departure of the Spanish community. I focus specifically on an intensive and heated debate that took place over the international Wikipedia list. Discussion lists are one of the main ways that participants in distributed projects remain in contact, debate and discuss issues, and identify as part of a community. They are a treasure trove of archival material for researchers of network cultures. While the Wikipedia discussion list considered in this chapter by no means represents everything that was said or written about the disagreements that led to the fork, it nonetheless represents a special kind of discourse. Discussion lists are by definition a form of public discourse. Because of this, and even though the tone on these lists is often very casual, statements expressed on lists have a public nature. They are expressed with the full knowledge that they will be subject to the conditions of acceptability—what might also be called "norms"—of the Wikipedia formation.

A Clash of Statements

It all starts with a seemingly innocent statement: "Bomis might well start selling ads."[5] For Sanger, the statement is not at all controversial. He is much more concerned with his own departure and in making sure others don't take it as some kind of cue that the project is in trouble or that Sanger is a

---

4. E. Enyedy, "Good Luck with Your WikiPAIDia" (2002), http://osdir.com/ml/science.linguistics.wikipedia.international/2002–02/msg00038.html. This exchange is also covered briefly in Andrew Lih's *The Wikipedia Revolution* (2009, 136–38).

5. Sanger, "Announcement about My Involvement in Wikipedia."

"quitter." Its uncontroversial character is betrayed in the way it is delivered, nestled among other statements and very much an aside. "Moreover, Bomis might well start selling ads on Wikipedia sometime within the next few months, and revenue from those ads might make it possible for me to come back to my old job. That would be great."[6] The uncontroversial-ness of it can also be seen in the way it is used to point to future realities and away from itself: "Bomis might well start selling ads . . . ads might make it possible for me to come back to my old job." In the language of Latourian statements, there is a "positive modality" attached to the statement which directs the reader away from the statement's own "conditions of production" (Latour 1987, 23)—an act that simultaneously makes the statement more real and hence less controversial. Because Sanger moves straight to the possibility of getting his job back, the reader is not expected to dwell on his statement about ads. The fate of this statement, however, was anything but uncontroversial.

Enter the dissenter: "Good luck with your wikiPAIDia." Edgar Enyedy is clear about what he is responding to. As is customary on discussion lists, he copies relevant sections of the last post (Sanger's) and pastes them into his own. This refreshes readers' memories but is also somewhat of a rhetorical device: it extracts the statement from its companions and its harmonious setting and places it under the microscope. The statement is bare for all to see. Underneath this repositioned, now-naked statement, Enyedy delivers a series of counterstatements: "Nobody is going to make even a simple buck placing ads on my work"; "I release my work in terms of free, both word senses"; "Nobody is going to use my efforts to pay wages or maintain servers." Enyedy is a native Spanish speaker, and his English requires a little extra effort, but his intention is clear enough: "Wikipedia should not have ads!" Far from being an uncontroversial statement, a mere aside, Sanger's statement is now hotly contested. For Enyedy, it has become a central matter of concern.

This is not the first time we have witnessed conflicting statements. I considered them in chapter 2 in regard to creating, editing, and writing article entries. We saw how these statements took part in "statement games" or "frame contests," negotiations of the rules, norms, and policies of Wikipedia. These battles were about what should be included in the encyclopedia (and hence only secondarily about what the encyclopedia "is"). There were pre-existing limits upon what kind of "moves" could be made to make a statement weak or strong, resilient or deletable, and I detailed these in chapter 3. The statements considered in this chapter are somewhat different. Without getting too far ahead of myself, the statements considered here are first and foremost

---

6. Ibid.

about what Wikipedia "is." They operate on a level that can radically alter the whole project (statement formation) and, as we shall see, the kinds of tactical "moves" available to contestants are more numerous (though not infinite).

Before getting into the thick of this controversy, I want to show something of the force of statements, and how they work to build and reduce realities, by lingering for a moment on Enyedy's initial response to Sanger. For clarity, I reproduce the entire post again and attribute a number to each statement I want to consider:

Larry Sanger wrote:

>[1] Bomis might well start selling ads on Wikipedia sometime within the
>next few months, and revenue from those ads might make it possible for me
>to come back to my old job.

Hi,

[2] I've read the above and I'm still astonished. [3] Nobody is going to make even a simple buck placing ads on my work, which is clearly intended for community, moreover, [4] I release my work in terms of free, both word senses, I and [sic] want to remain that way. [5] Nobody is going to use my efforts to pay wages and or maintain servers.

[6] And I'm not the only one who feels this way.

[7] I've left the project.

[8] You can see the Spanish Wikipedia development in the last two days and then you may think it over.

Good luck with your wikiPAIDia
Edgar Enyedy
[9] Spanish Wikipedia

After a brief greeting, "Hi," Enyedy wastes no time. The first part of statement (2) directs the reader to Sanger's earlier statement (1), "the above," while the second part aims to transform it. If the second part of Sanger's initial statement, "ads might make it possible for me to come back to my old job," operates as a positive modality, the second part of Enyedy's is a "negative modality." Of these Latour writes, "We will call *negative modalities* those sentences that lead a statement in the other direction towards its conditions of production and that explain in detail why it is solid or weak instead of using it to render some other consequences more necessary" (1987, 23). Enyedy is not only "astonished," but "still astonished," indicating his reaction had a certain durability to it. The modality is intended to weaken Sanger's initial statement.

When Latour writes about "negative modalities" in the formal and disciplined literatures of science and technology, they appear as "additions" to an initial, for convenience let's say "primary" statement. He writes: "We call these sentences *modalities* because they modify (or qualify) another one" (1987, 22). Although, as we have seen, Latour goes on to greatly expand what is considered a statement in other writings (an expansion that is already present in *Science in Action*),[7] the idea of modalities doesn't share the same fate. Indeed, the expanded notion of statements I have embraced thus far has relied on Latour's alternative but related notion of "loading." A statement is more or less loaded; it has more or less allies, more or less materiality. Modalities are a kind of literary or rhetorical equivalent, a kind of loading that exists only in language.

However, if a negative modality is anything that draws attention to a statement whose previous existence seemed settled, then statement (1) must also be considered a negative modality, even though this doesn't fit easily into Latour's schema. That is, the same words put forth by Sanger (re)appear in Enyedy's post but as a negative modality, or at least as *part* of a negative modality, because the very act of pasting it at the top suggests it is up for scrutiny. As Foucault reminds us, if "an identical formulation reappears, with the same words, substantially the same names—in fact, exactly the same sentence—it is not necessarily the same statement" (1972, 101). This is the case with statement (1), as the very same words have reappeared to work against and draw attention to their prior iteration. These words have transformed from statement to modality or perhaps from statement to "modal statement." However, it is complicated further still as the modality of state-

7. There is a passage in the middle of the text, for example, where Latour likens the statement to a ball in a game of rugby: "The total movement of the ball, of a statement, of an artefact" (1987, 104). Latour suggests that the ball metaphor is limited because the ball appears to be stable as it moves from hand to hand—it is "transmitted"—whereas statements are constantly modified as they circulate. However, if we take the ontology of ANT to its extreme, to its "forensic" limits, the ball too is constantly being transformed in subtle ways, even while maintaining its "formal equivalence." Not only is it constantly being marked by small abrasions, which slowly wear it down over time. The ball is subject to changes in trajectory and speed; it is allied with different players, who each do different things with the ball, things that the ball couldn't do with other players; it experiences changes in pressure and shape as it is kicked or hit forcefully. At the end of the game it is customary in many sports to give the ball to the best player, and so begins the ball's life as a trophy in a cabinet. While balls are well-built, sturdy things, they too are no less subject to modifications. It is also worth mentioning that Michel Serres uses this same example (it is not simply a metaphor) in his important work, *The Parasite* (2007, 225–27). It is from this example of the ball, circulating from hand to hand among players in a team, that Serres' theory of the quasi-object comes to life, breaking down the binary distinction between subject and object, and which forms a key philosophical building block of what would come to be known as ANT. On the reliance of the ball on the network of relations it enters into, Serres writes, "[a ball] is what it is only if a subject holds it. Over there, on the ground, it is nothing; it is stupid; it has no meaning, no function, and no value" (225).

ment (1) is not given in its entirety. The reader knows to expect something, the genre of discussion lists lets them know some kind of response to these words is coming, but statement (1)'s existence as a negative modality is only made clear or *completed* by statement (2). Statements (1) and (2) therefore work together: Statement (2) explicitly points to (1), "I've read the above," but really it isn't pointing to (1) at all. Instead, (1) is *standing in* for Sanger's initial statement when "pointed to" by statement (2). Statement (1) also simultaneously exists as its own unique statement, whose function as a negative modality is deferred and completed by statement (2). Without dwelling too long on these micro-complexities, we can nonetheless see that the status of statements and their modalities is often even messier than found in Latour's studies of technoscience. Much attention has to be paid not only to modalities within statements, but also to the modalities that exist between statements, in relation to one another and across sentences and even entire texts. We may speak therefore, of a *negative modal relation* set up between Sanger's statement and Enyedy's response(s).

The next, rather long and untidy, sentence is best considered in parts. It begins with (3), the first direct counterstatement: "Nobody is going to make even a simple buck placing ads on my work." It is likely that Enyedy intended to write "single buck," but it makes little difference. If statements (1) and (2) are negative modalities, designed to draw attention to and weaken Sanger's statement, statement (3) asserts its own reality, or rather two distinct realities: no money is to be made and Enyedy owns his work. The implication is that because Enyedy owns his work, he has control over its fate. He confirms this by adding a positive modality: his work is "clearly intended for community." Nobody is making money, because he owns his work, which is intended for the community. The force of the counterstatement is that it mobilizes the question of property (ownership of the work) and attributes it to Enyedy, and it also makes "community" stand in opposition to commercialization. If Sanger wishes to engage with this particular statement (which is not always the case in such debates), he must unravel these connections.

Having established (or at least attempted to establish) ownership of his contributions, in statement (4) Enyedy makes a gesture to free software: "I release my work in terms of free, both word senses." The "senses" he invokes refers to the "freedom" to edit the "source content" as well as free in the sense of price, or as is commonly remarked: "free as in free beer." Interestingly, although the gesture to free software is clear and must be seen as an attempt to place a powerful ally behind not having ads, it is actually a weak counterstatement. As we saw in chapter 1, free software is primarily about access to the code and not about price. Indeed, the founder of the Free Software

Movement, Richard Stallman, is not opposed to generating an income from one's outputs. Thus, while statement (4) attempts to bring in the principles of freedom derived from Free Software, and thus simultaneously make Enyedy a representative of this tradition, the way he invokes these principles will leave him open to attack as the debate plays out. The last part of statement (4), "[and I] want it to stay that way" is yet another positive modality. Statement (5) continues in much the same vein as (3) and (4), but adds a dimension of exploitation.

Statement (6) is a different kind of statement. To be sure, it participates in the *negative modal relation*, adding weight to the other statements, but it also does something more. Or rather, it adds weight to these other statements *because* it does something more. The statement "I'm not the only one who feels this way" suggests that Enyedy has allies, that there are other, unheard voices who support his counterstatements and the negative modal relation he has established in relation to a possible future with ads. This gesture situates Enyedy in a way that resonates with the notion of a "spokesperson," a concept that features prominently in the literature of actor–network theory (Akrich, Callon, and Latour 2002; Callon 1986; Latour 1987, 70–74; 1996, 42; 2005, 31–32). As the term suggests, a spokesperson is one who represents others, others who cannot or do not speak for themselves. The term is also explicitly directed to move beyond human-only forms of political representation. Thus, in Michel Callon's well-known account of the scallops of St. Brieuc Bay, it is not only "fishermen" but also the "scallops" that are spoken for, and Latour similarly writes not only of "Bill" who represents a group of workers who desire a pay rise, but also of "Davis" who speaks for "neutrinos that cannot talk" (Callon 1986; Latour 1987, 72).[8]

A spokesperson isn't just anyone who speaks for others. Rather, a series of negotiations, trials, and possibly even formal political mechanisms like elections must first take place to establish the spokesperson's representative character and to guarantee that the represented behave as the spokesperson claims. A voice can be deemed a spokesperson only when "the solidity of what the representative says is directly supported by the silent but eloquent presence of the represented" (Latour 1987, 72–73). There must be alignment between statements and the realities of those represented. The most important aspect of the spokesperson, however, is not whether or not that person is representative, or the fact that the person is representative of a posthuman

---

8. This is not to suggest that things without a voice do not *inscribe* the world in other ways.

political constituency; it is that the status of spokesperson imbues the representative's voice, his or her statements, with a special kind of force.

Callon makes explicit this connection between the (intermediaries) represented and the resulting force of the spokesperson's statements:[9]

> These chains of intermediaries which result in a sole and ultimate spokesperson can be described as the progressive mobilization of actors who render the following propositions credible and indisputable by forming alliances and acting as a unit of force: "Pecten maximus anchors" and "the fishermen want to restock the bay." (Callon 1986, 216)

For Callon, the statements "Pecten maximus anchors"—which refers to a type of scallop larvae and their ability to "anchor themselves to collectors and grow undisturbed while sheltered from predators" (1986, 203)—and "the fishermen want to restock the bay" become "credible" and "indisputable" because they are aligned with the realities they represent. Statements by spokespersons are loaded by way of *enrolling allies* who may or may not be explicitly manifested in discourse but are located elsewhere, standing in reserve. The force doesn't derive from actually loading the statement with materiality (as was the case with the speed bump or the weight attached to the hotel key), but simply by pointing to the realities spoken on behalf of. If these realities are secure and predictable, if they act as the spokesperson says, the spokesperson appears as a "mouthpiece" for a whole crowd—they speak for many and their statements have the force of many.

How does this relate to Enyedy and statement (6): "I'm not the only one who feels this way"? Is Enyedy a spokesperson, representing an entourage of dissenters? One thing that is clear is that statement (6), taken together with (7), (8), and (9), is designed to function in the same way as the statements that Latour and Callon attribute to those made by a spokesperson. Statement (6) invokes a crowd, "I'm not the only one," while (9) gives the crowd a name. There are, however, notable differences between Enyedy and the figure of the spokesperson.

First of all, spokespersons speak for those who cannot, either because those represented cannot speak at all, or they cannot speak under the particular conditions of statement production. (For example, Latour's character

---

9. While this is often referred to as a founding text of actor–network theory, I would like to emphasize that Callon explicitly saw his "sociology of translation" as "a new approach to the study of power" and "the role played by science and technology in structuring power relationships" (1986, 196–97).

"Bill" speaks for a body of workers, who all have the capacity to speak, but not all at the same time.) In Wikipedia, representative forms of decision-making are generally discouraged in favor of trying to achieve "consensus."[10] As we have seen in previous chapters, discourses of collaboration, consensus, and participation rule the day. In projects where everyone can speak, what role could there be for spokespersons?

Along with Wales and other official and unofficial representatives of and within Wikipedia, Enyedy is well aware of this predicament and responds to it in novel ways. First and foremost, Enyedy plays down any notion of formal leadership. When asked in hindsight about whether he led the revolt, for example, he refers to himself vaguely as "some sort of unofficial leader" and immediately names another important character, Javier de la Cueva. This remark is followed by a *pointing* back to those represented: "others shared our opinions" and "I did . . . receive a lot of support from the community" (Enyedy and Tkacz 2011, 115). Statement (6) is therefore well designed as it mobilizes the crowd while leaving Enyedy's status ambiguous. The challenge for Enyedy is to *be representative* without being *a representative*, to speak *the consensus*.

In his discussion of spokespersons, Latour also uses the notion of the "mouthpiece." The two terms are closely related and even used interchangeably, but I want to introduce a distinction to help describe Enyedy's ambiguous position in relation to statement (6). A spokesperson is first and foremost a role. It is a position that permits a certain force when producing statements afforded by that position. A spokesperson can also fail or misuse that role. For Latour, the figure of the spokesperson is ambiguous. Are spokespersons speaking for themselves or for the group? It is a question of agency. Do spokespersons put forth their individual desires, or are they more akin to a medium, letting others speak through their voice? Are they mediator or intermediary? In questioning whether a spokesperson is really a spokesperson, which is a question of whether or not that person speaks for him- or herself or for others, Latour uses the term "mouthpiece" to refer to those who actually do speak the voice of the many. Whereas the term "spokesperson" connotes not only representation, but also a certain agency of the person speaking, the term "mouthpiece" emphasizes the opposite. A "mouthpiece" is a passive medium, an intermediary, passing on statements whose force is derived from elsewhere. As a mouthpiece, Enyedy can speak with force, or

---

10. As David D. Clark famously remarked at the twenty-fourth Internet Engineering Task Force conference: "We reject: kings, presidents and voting. We believe in: rough consensus and running code" (Clark 1992, 543).

at least lay claim to such force, while avoiding being accused of speaking for others. Just as ad-hocracy is a mode of organizing that wants to avoid the very question of organization, the figure of the mouthpiece is a type of leadership that wants to deny the question of leadership. In open projects the figure of the mouthpiece, and its concomitant denial of the agency of the speaker, has become a common argumentative technique.

However, because the mouthpiece is not the source of the forceful statements the mouthpiece produces, because he or she makes a point of eschewing authority, there is even more pressure to refer back to those the mouthpiece speaks *with*. The mouthpiece is under constant pressure to point to his or her crowd and also to defend against the possibility of being accused of speaking for (as opposed to with) others. Enyedy responds to this predicament in statements (7), "I've left the project," and (8), "You can see the Spanish Wikipedia development in the last two days and then you may think it over." Statement (6) sets Enyedy up as a mouthpiece, thereby increasing the force of the negative modal relation. Statement (7) indicates a course of action that he has taken in response to Sanger's initial statement and statement (8) is designed to demonstrate the alignment of Enyedy's actions (and his statements) with the rest of the Spanish Wikipedia contributors by pointing to an observable reality (the lack of activity on the Spanish Wikipedia).

Interestingly, while Enyedy uses the technique of the mouthpiece in the post above, in other writings he presents himself more like a spokesperson:

> I recognised that people wanted to make suggestions, to debate and be heard. But those kinds of processes can be lengthy, so I made the decisions. I thought the timing was critical—a line had been crossed and I didn't want it to be a never-ending story. Luckily the community supported me. This was the extent of the unofficial leadership. I made a decision and others supported it. (Enyedy and Tkacz 2011, 115)

In this passage, Enyedy is writing about the general decision to fork and is more willing to assert his agency, perhaps because his (individual) actions (covered in the previous chapter) are hard to allocate to the crowd. Note still, though, his reluctance to claim that he is speaking and acting on behalf of anyone: Enyedy was "lucky" his decisions were supported. The decision to take action is also justified by the importance of "timing" and the fact that "a line had been crossed," which represent alternative ways of playing down his own agency. Indeed, much like Latour's character "Bill" who represents the workers, Enyedy recognizes that others can speak, "to make suggestions, to debate and be heard." While Bill's workers could speak, but not all at once, here it is time that acts as the constraint.

Enyedy's short post, "Good luck with your wikiPAIDia," reveals much about the politics of statements in open projects and in distinction to their origins in the science and technology studies of Latour and Callon. Enyedy provides several counterstatements; creates a generalized set of negative modal relations in response to Sanger; and sets himself up as a mouthpiece while fending off possible accusations of assumed authority and leadership, that is, of acting as the source of power. Statements, counterstatements, positive and negative modalities, and spokespersons are all in play. However, to account for the particularities of open projects and the "discussion post" genre in which the statements are deployed, I have added the notion of "modal relations" and placed new emphasis on the notion of the "mouthpiece." The negative modal relationship is established, even though the parameters of the debate are yet to be fully determined and what might be considered the primary counterstatement, "Wikipedia should not have ads!," is never actually articulated. The notion of negative modal relationships is therefore well suited to the messiness of debates, where the precise nature of the conflict is less certain than the mere fact of it, and when counterstatements are still being formulated. It is the negativity from which alternative realities emerge. I now follow the debate about ads as it passes through other voices, is modified, challenged, and disputed. I will try to stick as closely as possible only to the question of ads, although, as we will see, the question of ads always leads us in other directions, pointing to outside realities and future possibilities. The controversy about ads becomes nothing less than a full-fledged debate about Wikipedia—about which statements define the project as a whole.

Statements in Action

Sanger is quick to reply and makes a series of counterarguments. He begins by stressing how vital his paid employment has been to the different Wikipedias: "If I had not been paid to start them, they would not exist."[11] For Sanger, Wikipedia cannot thrive without paid staff members. Since Wales had run out of money for the project, ads presented a possible solution. The logic is simple:

1. Wikipedia needs paid staff to thrive.
2. Ads might generate enough money for paid staff.
3. Therefore, Wikipedia should have ads.

---

11. L. Sanger, "Re: Good luck with Your WikiPaidia" (2002), http://osdir.com/ml/science.linguistics.wikipedia.international/2002–02/msg00039.html.

The proof of statement one is seemingly assured by Wikipedia's very existence (as noted above) and its successful development up to this point:

> The benefits you have been experiencing on the Spanish Wikipedia—and there are many—are in large measure because Jimmy Wales has paid me and his programmers (including himself) a lot of money over the past two years getting all the infrastructure set up.[12]

If this argument holds, generating revenue to pay for staff will become an accepted part of the reality of Wikipedia, placing Enyedy and his motley crew of dissenters in a difficult position. But Sanger isn't finished building this point:

> There is a silly theory a lot of people have about nonprofit projects generally, namely, that the people who organize those projects SHOULDN'T BE PAID ANY MONEY. They should work for free, too. Well, if that happens, then the nonprofit projects in most cases JUST WON'T HAPPEN. Those who organize nonprofit projects are doing a very valuable service, one that in many cases just wouldn't be done if they weren't paid. I'm sorry if you don't like that, but people need money to survive. That's the way the world works.[13]

Now it is not only Wikipedia that benefits from paid staff but all nonprofit projects and to deny this would be "silly." To stand against Sanger is now to stand against all nonprofit projects generating revenue and is based on a profound misunderstanding of "the way the world works." Sanger is trying to build allies and connect his statements to well-established realities, like "people need to survive." From his initial statement about Wikipedia, Sanger has now significantly inflated his position; it is no longer a statement about Wikipedia, but how the world is.

In the final paragraph of his response, Sanger makes three further moves. First, he writes, "it has long been explicitly declared in several places that Wikipedia would EVENTUALLY runs ads that would pay for me" and points to the "FAQ" as evidence of this reality. Not only is the possibility of having ads totally acceptable, it is now given a long history within the project. Second, Sanger makes an intervention into the terms of debate. He turns the main actor in dispute, ads, into "sponsorships" that result from "donations," offering the rationale that "it is now nearly certain that Wikipedia will be a nonprofit" (at the time it still technically remained a commercial enterprise).[14] Enyedy might be against ads and people making money from his work, but

12. Ibid.
13. Ibid.
14. Ibid.

what about innocent contributions from benevolent philanthropists whose only intention is to keep this noble project running? Finally, Sanger distances the question of money from his own circumstances and tries to bring it closer to the realities of the Spanish contingency:

> I really hope that we eventually make *more* than needed to pay for me, so we can hire others to work full-time as well. For example, it's long been my opinion that it would be wonderful if we could hire a Spanish language Wikipedia organizer.[15]

Sanger's response is an attempt to counter the negative modal relation set up by Enyedy. The possibility of ads is connected to a thriving Wikipedia, the valuable work of nonprofits, and the realities of how the world works. Ads in Wikipedia are given a long history (implying that those unaware of this history are uninformed), transformed into sponsorships, distanced from Sanger's personal interest, and connected to a future of paid Spanish contributors. Having now realized the controversial nature of his initial statement, Sanger has mobilized his own set of (counter-counter-) statements in favor of ads. Although each writes in an assertive manner, the fate of these competing positions is at this stage far from certain.

From this point onward, new voices begin to emerge. The first was Tomasz Wegrzanowski, a prominent member of the Polish Wikipedia.[16] Wegrzanowski challenges Sanger's claims about the value of full-time employment: "they don't need anybody employed full-time to take care of them." To strengthen this statement, he proposes a different reality for Wikipedia, disconnecting it from the generalities of nonprofit organizations: "Wikipedias (communities) create content themselves and, to [a] big extent, manage themselves" and further adds, "If self-management won't work well with bigger Wikipedias (100k or even 1M articles), software can be changed to bring better solutions."[17] For Wegrzanowski, Wikipedia can take care of itself and any managerial complexity can be dealt with via software.

---

15. Ibid.

16. The Polish Wikipedia had its own unique tensions with the founders. The project began on a different server (wiki.rozeta.com.pl) to the other Wikipedias and although they had decided to join the other projects on the Bomis server, there was a period where they returned to the original Rozeta wiki. The controversy, which was largely over software updates (and specific issues relating to non–Latin language characters), is often described as the first fork of Wikipedia. This history, and especially the existence of the alternative wiki on the Rozeta server, would shape Tomasz's comments as well as their reception.

17. T. Wegrzanowski, "Re: Ads and the Future of Wikipedias" (2002c), accessed March 28, 2011, http://osdir.com/ml/science.linguistics.wikipedia.international/2002–02/msg00041.html.

Wegrzanowski also explicitly allies himself with Enyedy and against ads: "And Edgar is right that ads are bad. Not only they're annoying but what warranty do we have that Bomis won't advertise Microsoft software or something equally evil?" He urges Edgar to set up a *working model* comparable to Wikipedia so that the community is not bound to Bomis, and it is here that some of the arguments about the power of forking considered in the last chapter are mobilized: "If you are also worried that Bomis might start to behave irresponsibly some day, find someone willing to host your Wikipedia under better terms. It will either assure that Bomis will behave more reasonably, or if they won't, they will immediately lose."[18] Wegrzanowski finishes his post by providing advice to Edgar, other Wikipedians, and Bomis, which largely reiterates his main points as well as asking Bomis to "at least publish some reasonable ad policy so we know what are your plans."[19]

Sanger responds to Wegrzanowski by questioning his "adversarial stance" and reminding him that "We're working on this all together." Once again he challenges the stance against ads: "You both have this knee-jerk reaction to them and give no reasons to think they're bad. Ads support many, many worthwhile projects." He does, however, note that the details of how to implement ads are open for discussion and we begin to see the shape of his ideas: "contextually relevant, simple text ads . . . displayed to people who are not signed in."[20] While sharing these ideas, Sanger comes close to acting as a spokesperson:

> it's going to be VERY difficult to convince most people who work on the project to fork. I hope I'm not going out on a limb by saying that most people would be happy to let our readers endure some text ads if it would mean that I, and eventually hopefully some other well-qualified people, could work full time organizing and improving Wikipedia and Nupedia.[21]

While Sanger treads carefully, he nonetheless invokes a crowd, "most people," but unlike Enyedy he does not *point* to it; he does not attempt to be the mouthpiece.

Finally, while trying to deflect the focus on Wales's commercial enterprise, Sanger alters or in this case replaces another actor, Bomis, and simultaneously

---

18. While Wegrzanowski doesn't explicitly use the term *fork* here, responses to his comments describe his suggestion as forking. I do not cover the entire thread about forking, but from this point on in the debate the merits of forking are also considered.

19. Wegrzanowski, "Re: Ads and the Future of Wikipedias" (2002c)

20. L. Sanger, "Re: Ads and the Future of Wikipedias" (2002), http://osdir.com/ml/science.linguistics.wikipedia.international/2002–02/msg00041.html.

21. Ibid.

the future nature of the project: "And it might not be *Bomis* that rents the sponsorship space: it might be a nonprofit Nupedia Foundation. This has yet to be finalized."[22] He completes this vision of a future foundation by providing more details about paid staff:

> Five or ten full-time employees are REALLY, REALLY needed, if this is going to become a really world-class resource. There are many ways in which the projects, both Wikipedia and Nupedia, will stagnate and possibly go downhill if I am not constantly and actively involved; no one will volunteer to do a lot of the stuff I'm expected to do there [here?], and I know that because I've asked *many* times for help.[23]

Obviously, Sanger doesn't think that either software or volunteer labor can replace his function. As the controversy about ads develops, we see new actors (allies) brought into the mix (nonprofits, software, the world, the Rozeta wiki); current actors are altered in an attempt to reconcile the opposing positions (ads/sponsorships, Bomis/Nupedia Foundation); and future realities are invoked and challenged. Arguments are simultaneously becoming more general and more specific as the contours of the debate begin to emerge. It becomes difficult to focus only on ads because this controversy is itself connected to other things, to long chains of allies and complex modalities. The controversy over ads is getting larger, increasing in force as allies are built on both sides and resolution remains out of sight.

Wegrzanowski replies that although everyone is working together "on Wikipedia," Bomis makes decisions "without asking anyone."[24] His statements weaken Sanger's reminder that "We're working on this all together." Wegrzanowski also refutes Sanger's claim that most people would be okay with ads and provides a list of reasons why "ads are bad":

- they provide *no* useful information. if they did, it would be included in articles
- they leave crap in reader's mind
- they often promote evil things
- money from ads may make some people less objective.[25]

Wegrzanowski finishes by questioning Sanger's vision of the future and reiterating his earlier points about the organizational capacities of software: "Five

---

22. Ibid.
23. Ibid.
24. T. Wegrzanowski, "Re: Ads and the Future of Wikipedias" (2002b), accessed March 30, 2011, http://osdir.com/ml/science.linguistics.wikipedia.international/2002–02/msg00042.html.
25. Ibid.

to ten? What for? Wikipedia isn't going to stagnate or go downhill, but some serious enhancements in software are required to ease self-management."[26]

A new discussant, Joao Miranda, enters the controversy with a short post. Miranda makes a new claim about ads: "Wikipedia articles are free to everybody to use. So, if there is money to be made with ads, somebody will profit from your work. It can be Boomis [sic] or Yahoo or Microsoft."[27] Although Miranda's statements lend themselves to Sanger's position, their primary function is to lessen the force of the entire controversy. If there is no absolute way to disconnect Wikipedia articles from commercialization, indeed, if this possibility is inscribed in the very license adopted by all Wikipedia articles, why make such a fuss? And because of this fact, why shouldn't it be Bomis, which is at least committed to the project? Wegrzanowski concedes Miranda's point about the possible commercialization of content, but argues that what is important is the "availability of an ad-free version."[28] The mere possibility of ads and making money is not a reason for Bomis to implement them. Is the force of the debate restored, or is it weakened by Miranda's statements? Still the debate continues.

Jim Wales makes his first contribution. If Enyedy was "amazed" at the possibility of ads, Wales is similarly taken by the controversy itself: "Gee, what a strange bunch of messages." His post is very short, and besides this opening statement he only makes two further points: (1) he directs readers to a "several months old" statement he made about the possibility of ads on his personal Wikipedia page; and (2) he strongly states that he is not "going to do ANYTHING without asking people."[29] Perhaps sensing the building hostility toward Sanger, Wales adopts a different tone—the kind of well-intentioned, considered modesty that would come to make him one of the most celebrated "benevolent dictators" in open projects. Without going into too much detail at this point, in the posts that followed Wales allies himself with the possibility of incorporating ads, notes that the Free Software Movement has always been open to making money (refuting Enyedy's earlier invocation of this ally), critiques the logic that would enable some companies to "make money" from the project while eschewing any attempt to "raise money for the project," notes that the details of how to implement ads are open for

26. Ibid.
27. J. Miranda, "Re: Ads and the Future of Wikipedias" (2002), accessed from http://osdir.com/ml/science.linguistics.wikipedia.international/2002-02/msg00043.html.
28. T. Wegrzanowski, "Re: Ads and the Future of Wikipedias" (2002a), accessed March 29, 2011, http://osdir.com/ml/science.linguistics.wikipedia.international/2002-02/msg00044.html.
29. J. Wales, "The Future" (2002), accessed March 29, 2011, from http://osdir.com/ml/science.linguistics.wikipedia.international/2002-02/msg00045.html.

discussion, modifies the actor Bomis by noting, "It's just me. Talk to me. I'm a real human being who founded this project out of love for knowledge, love for freedom," stresses on multiple occasions that he is "totally open to input," claims that for the Spanish community to be on strike they must have been lied to, and accuses Enyedy of "total ignorance" for not bothering "to ask questions before attacking me [Wales] publicly."[30]

As the controversy enters its final stages, still with no clear resolution in sight, discussion of ads gets more and more technical, that is, more specific, more detailed, and thus larger and more general. Wales and other discussants consider what kind of ads would be acceptable, when ads would be implemented in the different Wikipedias, and roughly how much money they would generate. The negative modalities are directed less at ads specifically and increasingly at the way the project has been governed in general. Questions of representation, or spokespersons, and of allies increasingly become manifest.

The Dissenter, the Mouthpiece, and the Failed Spokesperson

In *Science in Action*, Latour introduces his readers to the fictional character of "the dissenter." This literary device enables Latour to follow the workings of science and technology as it builds its facts and technical devices by introducing radical doubt at every step. The dissenter is first and foremost a disbeliever of a statement-fact; it is a creator of controversies. The dissenter interrogates the statement under contention in a series of ways, beginning at the level of discourse, through to observation of the actual realities from which the statement is produced and finally, if they are able, by creating competing (counter)realities. Each of these moments interrogates one dimension of the *conditions of possibility* from which the statement emerges: "people in disagreement open more and more black boxes and are led further and further upstream, so to speak, into the conditions which produced the statements" (Latour 1987, 30).

At least in the first stages, the dissenter's investigation moves backwards. To challenge a statement is to look past it, by looking through it. If these conditions, these prior realities, can themselves be doubted or dismissed, the statement will most likely share the same fate. For example, earlier in the con-

---

30. J. Wales, "Re: Ads and the Future of Wikipedias" (2002), and "Re: Spanish Wikipedia on Strike" (2002), accessed March 29, 2011, from http://osdir.com/ml/science.linguistics.wikipedia.international/2002–02/msg00049.html and http://osdir.com/ml/science.linguistics.wikipedia.international/2002–02/msg00067.html.

troversy Sanger is asked to detail why several paid staff members are needed. If he can provide a list of tasks and simultaneously demonstrate why software or volunteers could not perform these tasks, his claim is more likely to be accepted and so too, therefore, is the need to generate some kind of revenue. That is, his statement will have pointed to a convincing reality and therefore gained a new ally: a list of activities that are uniquely done by paid staff.

Latour shows how the dissenter moves further and further into the conditions of possibility by way of a conversation about a new cure for "dwarfism":

> Mr Anybody (as if resuming an old dispute): "Since there is a new cure for dwarfism, how can you say this?"
> Mr Somebody: "A new cure? How do you know? You just made it up."
> "I read it in a magazine."
> "Come on! I suppose it was a colour supplement . . ."
> "No, it was in *The Times* and the man who wrote it was not a journalist but someone with a doctorate."
> "What does that mean? He was probably some unemployed physicist who does not know the difference between RNA and DNA."
> "But he was referring to a paper published in *Nature* by the Nobel Prize winner Andrew Schally and six of his colleagues, a big study, financed by all sorts of big institutions, the National Institute of Health, the National Science Foundation, which told what the sequence of a hormone was that releases growth hormone. Doesn't that mean something?"
> "Oh! You should have said so first . . . that's quite different. Yes, I guess it does." (Latour 1987, 31)

On this occasion, the dissenter (Mr Somebody) is unsuccessful. The initial statement, "there is a new cure for dwarfism," has garnered many allies, too many for Mr Somebody to doubt. For Latour, Mr Anybody has become "Mr Manybodies" (31); he speaks with a crowd behind him.

However, in Latour's examples, if the dissenter is still not satisfied with the allies mobilized in discursive form, she can look to the actual realities invoked. With Latour's example of dwarfism, this would involve going into the laboratory where the study was conducted and seeing the sequence of this hormone for oneself. Finally, if the dissenter is still not satisfied, if she is not convinced by what she has seen in the laboratory, she can try to create competing realities. It is in this third moment that all the old power dynamics identified by Marxist-inspired investigations of science and technology come into play. Not just anyone can create competing realities from which to produce competing statements of comparable pedigree. To create these realities, one must be able to mobilize an army of allies. For Mr Somebody, this includes national institutes and foundations, expensive laboratories, high-profile colleagues, prestigious journals, and so on. Only with all these realities, these conditions of possibility, securely in place, could Mr Somebody produce counterstatements about hormone sequences.

Latour does not follow the dissenter through these various phases of technoscientific practice in order to dismiss them as ideological, or even to show the social or human (and hence "flawed") aspect of hard science more generally. Rather, the dissenter is invoked to show the durability of well-constructed statements. It is an investigation into how some statements and their realities stick and gain traction while others don't. The dissenter is a devil's advocate, noise in the system, which shows the positive function of statement construction. The dissenter makes sure the scientist "speaks for" all allies drawn upon, that the statement is aligned with its conditions of possibility, its hinterlands. It is a question of representation. In the first two phases the dissenter challenges the representativeness of the speaker and his statement, while in the third phase the dissenter must become a spokesperson. In reality, of course, these phases are not neatly delineated and can proceed in any order and combination.

Who or what is the dissenter in our controversy? Is it Enyedy? Is it one, or perhaps several, of his statements? Does Sanger also become a dissenter, a counter-dissenter? What about Wegrzanowski and his statements? Earlier, and with deliberate ambiguity I described "Good luck with your wikiPAIDia" as the dissenter for reasons I will now explicate. The dissenter is defined primarily by what it does. What or who *it is*, its identity, is nothing more than this *what it does*. As I have already outlined, what the dissenter does is contest statements; it creates controversies and challenges realities where previously there was harmony. Following Chantal Mouffe, we might say the dissenter is the mechanism that transforms what she describes as *the social* into *the political*. It is clear, then, that by itself a statement is not a dissenter. There are many contradictory statements that coexist in relative harmony. It is not until such statements are juxtaposed, not until they are brought into relation, that controversy begins. What about negative modalities? A negative modal relation can be present before any counterstatement is positively stated. And, indeed, by challenging the reality of a statement and pointing to its conditions of possibility, negative modal relations are certainly expressions of dissent. The dissenter is therefore a bringer-forth of negative modalities. But the dissenter's activities extend beyond creating negative modal relations.

Is the dissenter an author or author function? A subject position? Enyedy authored the original "wikiPAIDia" post; he brought forth the negative modal relation and juxtaposed Sanger's statements with his own counterstatements, but he didn't act alone. Moreover, the dissenter not only brings forth, but also *persists* until the controversy is settled (even if this settlement is only temporary) and new relations are established. Because of this, Wegrzanowski and other discussants must also be included. The dissenter is plural.

The dissenter must also bring in allies. Indeed, the dissenter is nothing without allies and it is these same allies that make the dissent both possible and unavoidable. The dissenter is caught between two contradictory realities, two sets of statements, even if only one (the one being challenged) is made explicit. To persist with the dissent, the dissenter must eventually refer to these realities (and, as we saw, possibly even create them). Even in the short example of dwarfism considered above, the dissenter brings in allies (counter-realities): the color supplement and the unemployed physicist who doesn't know the difference between RNA and DNA. Like many characters in the ontology of actor–network theory, the dissenter thus becomes a network, an assemblage of dissent, acting more or less in unison (the extent of which will be tested)—juxtaposing counterstatements, attaching negative modalities, and creating negative modal relations. The dissenter sways between (quasi-) subject position and assemblage. Thus, representation once more becomes important. The dissenter is subject to the same representational conditions as the spokesperson. The dissenter must become a spokesperson, must speak for (or "with") the counter-crowd, the "dissent assemblage." "Good luck with your wikiPAIDia" is the name I have given to the dissent assemblage that mobilizes around Sanger's statement. Enyedy is a crucial part of this; he is a dissenter, to be sure, but he does not and cannot act alone.

I have already described the specific technique of the successful spokesperson in open projects as that of the "mouthpiece." As the controversy neared its climax, techniques of representation became increasingly significant. By this stage, allies have been amassed on both sides and it seems that there are too many counterstatements to resolve. Although by no means acting alone, Enyedy has become the chief spokesperson of dissent, with Sanger and to a lesser extent Wales assuming the same role on the other side. As I noted earlier, these two figures are different kinds of representatives, with Enyedy striving to be a "mouthpiece" and Sanger a "spokesperson." Beginning with Wegrzanowski, Sanger (acting on behalf of Bomis) is accused of not representing the crowd: "We work together on Wikipedia, but Bomis does such things without asking anyone."[31] Following this, Enyedy makes a series of similar remarks in a lengthy post, designed to cut the link between the current

---

31. Wegrzanowski, "Re: Ads and the Future of Wikipedias" (2002b).

leaders and their constituents, such as, "We are a dot.com behaving like a dot.org . . . It's a double-faced game" and "Shouldn't wikipedians be asked first?"[32] The first two statements separate the intentions of the founders (Wales, Sanger, Bomis) from that of the community. These are followed by a direct challenge on the issue of representation. In the same post, Enyedy continues to foster his status as "mouthpiece." Here are some examples:

> I've gathered many complaints . . .
> They feel their contributions are being kept by Bomis in some way. And they have the right to move if they want to, but they feel they can't.
> And they are planning to move or whatever they are planning without my guidance, I'm only receiving mails, but I haven't answered any of them yet. I said, 'I've left' and that's what I behave. Many of them have told me about such a strike until they get clear proposals, and you can see Recent Changes on Spanish Wikipedia whenever you want.[33]

Note the kinds of agency Enyedy grants himself: a gatherer, receiver, listener—agencies of the mouthpiece. When he refers to his own departure, he does not speak of or for others, only of the equivalence between his statement and his action. The community is depicted as completely autonomous, operating "without guidance" and once again, he points to their reality—the lack of "Recent Changes" on the Spanish Wikipedia—to shore up his own statements. Enyedy finishes by bringing these two tactics of representation together:

> It seems that Wikipedia is ruled by elitism, the most powerful members of this community are able (and have the rights) to delete pages, to place ads, and to decide what's the best for us. Surprisingly, we also have the ability to decide, despite being the "outer members from small wikipedias."[34]

In replying to Enyedy, Wales too tries to act like a mouthpiece. He writes:

> If it seems that way to you, you've obviously never written to me personally to voice your concerns.
> I'm totally open to input from all sides as to the best way to proceed. The degree of paranoia that I see sometimes is totally unjustified.
> One rhetorical "angle" that I do not like is for people to refer to "Bomis" as if it were some abstract entity. It's just me. Talk to me. I'm a real human being who founded this project out of love for knowledge, love for freedom.

---

32. E. Enyedy, "Ads and the Future of Wikipedias" (2002), accessed April 7, 2011, from http://osdir.com/ml/science.linguistics.wikipedia.international/2002–02/msg00048.html.
33. Ibid.
34. Ibid.

To accuse me of some kind of corruption is just too much—and not likely to get you what you want.³⁵

Speaking as a mouthpiece is here associated with being "totally open," even though Wales goes on to complicate this logic by acting as gatekeeper: you are "not likely to get what you want." Note also how Wales is well aware of the statement games in play around the invocation of Bomis as "some abstract entity," although this phrase is a pretty accurate general description of all companies. Wales responds to this "rhetoric" by trying to humanize and concretize Bomis, make it less abstract and more real, by associating it with himself.

While Wales and Enyedy adopt similar approaches, Sanger's tone, that of the spokesperson, is quite different. He does not hesitate to emphasize his special status within the project—for example, as "THE VERY PEOPLE WHO STARTED THESE PROJECTS"—or to assert behavioral protocols—"the proper response, when someone feels insulted at something you've said, when you didn't intend it, is: apologize." He also doesn't hesitate to insult the opposing side's position: "Anyone who says that paid employees are useless and not needed just doesn't have a clue what he's talking about," and after listing his (paid) duties, "If you think this can all get done with volunteers—my reply is, BWAHAHAHAHAHAHAHAHAHAHA!!!!" Indeed, even when Sanger attempts to be a good spokesperson, he cannot relinquish his own agency: "the management of a nonprofit Nupedia Foundation would be very responsive to the participants of Wikipedia (and Nupedia). I would see to it that this is the case."³⁶ If Enyedy tries not to be a leader at all, Sanger is willing to personally ensure that "management" is "responsive."³⁷

Toward the very end of the controversy and perhaps sensing that he is *outnumbered*, Sanger pleads for new allies:

35. J. Wales, "Re: Ads and the future of Wikipedias" (2002), accessed April 11, 2011, from http://osdir.com/ml/science.linguistics.wikipedia.international/2002–02/msg00049.html.
36. L. Sanger, "Re: Five Messages" (2002), accessed April 11, 2011, from http://osdir.com/ml/science.linguistics.wikipedia.international/2002–02/msg00060.html.
37. At this point it's worth mentioning Sanger's own pseudo-fork of Wikipedia, Citizendium. Sanger announced the Citizendium project in 2006, which was designed to correct what he perceived to be the main flaws with Wikipedia. The main thrust of Citizendium was to reintroduce the idea of expertise and more traditional mechanisms of quality control into the project. Sanger sought to connect Citizendium to the "outside world" in ways that Wikipedia never did (by requesting real names, age limits, university qualifications, and so on). For example, the rough equivalent of an Administrator (the Wikipedia user access level) in Citizendium is a Constable, who is equally tasked with "maintaining" the site and "enforcing" the rules, but must also generally be over twenty-five and hold a university degree. This future fork is worth mentioning precisely because one can already detect, in the EL fork controversy, aspects of Sanger's views about how best to organize knowledge, and specifically in relation to expertise.

non-critics, if you support my cause and think you can supply the correct response to an e-mail that, you can predict, I will want to have answered, please do that. Don't hold back and wait for me to do it. I'll be very grateful.[38]

And a few heated posts later: "Folks, I could use some expression of support here. I really DON'T think I should be expected to put up with this sort of treatment constantly and with a smile."[39] Once again, though, even when Sanger is pleading for help, when he is trying to muster a crowd, and when it seems like he has been abandoned, he continues to assert his agency *apart* from the crowd: "if you support *my cause*" and predict what "*I will want.*" Only one person came to Sanger's defense and a week later the list was informed that the Spanish community had forked.

Controversies are constituted by conflicting modalities, statements and counterstatements, dissenters and spokespersons, and their respective assemblages. To follow a controversy is to give life to all these characters and elements. We have seen, however, that not all dissenters and spokespersons are created equal. Indeed, asymmetry is the goal of the dissenter-spokesperson: to mobilize a larger, more forceful set of allies than the opposition. On top of this, there are different techniques of representation and these affect the type and number of crowd one is able to gather. As a controversy moves from building ally-assemblages to testing the representative character of the spokesperson in relation to that person's assemblages, these techniques of representation become increasingly important. In the end, Enyedy's technique of the mouthpiece-dissenter was successful. He was able to produce powerful statements for a whole crowd without placing himself in any position of authority. Enyedy could express authority while claiming not to be its source. Sanger, on the other hand, was a failed spokesperson. At the end of the controversy it is as if he turned around and no one or thing was behind him. This does not mean his position was wrong from the beginning, or wrong at all for that matter. Rather, it means that he was unable to convince a crowd that he was correct; he was not able to point to convincing realities and thus he was not able to produce new realities. While Wales was not able to turn the tables in this controversy (after stepping in late), he too adopted the technique of the mouthpiece, although we saw this weakened by the existence of Bomis.

When a project forks, it is common for people to put the reasons down to "personal disagreements" or "character clashes." It is at best convenient

38. L. Sanger, "Plea for Help" (2002), http://osdir.com/ml/science.linguistics.wikipedia.international/2002–02/msg00068.html.

39. Sanger, "Re: Five Messages."

shorthand and at worst yet another way to make politics invisible. Through the conceptual tools I have developed, however, it is possible to see the function of the becoming-personal of controversies. It is by focusing on the individual, the spokesperson, that the unity of the assemblage is tested. In Latour's words, it is a trial of strength. By attacking the spokesperson, dissenters try to create a condition of nonequivalence between this figure and the crowd. Does this figure really speak for all assembled, or merely for themselves? In the final section of this chapter I want to reflect more generally on what forking reveals about force in statement formations.

Forking and Statement Formations

Throughout this book, I have approached Wikipedia as a statement formation. In previous chapters I have shown how competing statements battle for existence within the rules of the formation, rules which themselves are understood as especially forceful statements (such as the neutral point of view). I have also stressed the performative nature of statements, how they attain force and their differing materialities. Forking represents a unique vantage point from which to attend to statements and their formations. When the possibility of a fork emerges, the controversy cannot be settled within the current rules of the formation.[40] The threat of forking is a challenge to the formation (and its rules) itself; it is a contest over the most forceful statements within a formation, over statements that are *definitive* of the formation. However, forking doesn't merely make visible pre-existing, already powerful, conflicting statements. Rather, the force of a statement is only attained *through* the controversy. Statements about the commercial status of Wikipedia, for example, only became powerful after allies were assembled and adversaries overcome. Only then did "Wikipedia will not have ads" move beyond contestation. (To be sure, if left unchallenged Sanger's initial statement might also have become very influential and, after a certain duration, heavily loaded and thus difficult to separate from the formation. Controversies are by no means the only way to build allies and become *definitive*.)

Like all forms of political exit, forking is usually viewed as a last resort. While in the previous chapter I stressed the perceived lossless quality of forking, I also pointed out how the current literature does recognize loss in terms of the community or the project's "symbolic aspects," for example. Forking may not be as destructive or politically risky as other forms of exit, but it is

---

40. This does not mean that the controversy is free to go in any direction. All statements and all controversies must draw on hinterlands, which both limit and enable what is possible.

perceived to have a destabilizing effect. This is because the two projects most often end up in direct competition, with one project becoming more successful (more real). Alongside this destabilizing effect, I want to suggest that the exact opposite is also true: forking is a process of stabilization; it is a making-durable of statements that become *definitive* of an open project. There is no contradiction here. Forking destabilizes on several fronts: it removes energies and allies from the original project; it challenges the forkers to create a *working model* of formal equivalence to the original; and eventually, as was the case with the EL and the Spanish Wikipedia, one project will emerge as more real than the other. At the same time, however, the statements that organize a project or two competing projects, will, as a result of the fork, become heavily loaded and thus more certain. Even though the EL failed, "Wikipedia will not have ads" was a resounding success. And it was not only this statement that became definitive.

Forking forces a general process of definition. As our controversy developed and allies amassed, it was not possible to contain the disagreement to the question of ads. The contours of debate became aligned with the contours of the project. The whole formation was up for grabs. Will Wikipedia be overseen by a foundation? What will the governance structure look like? Will there be paid employees? How many? Upon reflecting on the fork, Enyedy writes:

> I would like to remark upon the fact that as it is known today, the international Wikipedia that you all know and have come to take for granted, might have been impossible without the Spanish fork. Wales was worried that other foreign communities would follow our fork. He learnt from us what to do and what not to do. The guidelines were clear: update the database; make the software easily available on Sourceforge; no advertising at all; set up a foundation with a dot-org domain and workers chosen from the community; no more Sanger-like figures, as well as some minor things I haven't mentioned, such as free (non proprietary) formats for images. (Enyedy and Tkacz 2011, 115)

Enyedy sees the Spanish fork as producing "clear guidelines," all of which were based on issues that intensified during the controversy. He and the other dissenters forced the future direction of Wikipedia.[41] Wales's take on the event, however, is completely different: "The Spanish fork did not provoke any changes of any kind. We stayed the course. I didn't want to have advertising, and I found ways to avoid it—the Spanish fork was an important event in the history of Wikipedia, but not in the sense of 'provoking change'"

---

41. Equally, while Sanger "lost" the controversy, his countervision would largely be realized in the Citizendium project.

(Wales, in Tkacz 2011). Wales is understandably reluctant to recognize the role of the fork within the mythologized history of Wikipedia, especially because not implementing ads would become a crucial identifier of Wikipedia and would eventually be linked to him as the source. Sanger's take is different again, and closer to what I have argued: "But to give credit where it is due, Mr. Enyedy is correct that the fork of the Spanish Wikipedia might well have been the straw that finally tipped the scales in favor of a 100% ad-free Wikipedia" (Sanger, in Tkacz 2011). Although, rather than seeing the fork as the straw, the fork is actually more akin to the scales: it is about adding weight, juxtaposition, and weighing up.

Placing emphasis on the stabilizing function of forking also permits a reading of the Spanish fork as far more significant in terms of the original project. In the years that followed, Wikipedia would indeed implement all of the "guidelines" listed by Enyedy. This perspective on forking is also supported by Enyedy's own view of the event:

> The fork had its time and place, its goal and its consequences. Nowadays, the romantic point of view is that EL survived and is still going strong. It is a nice view, but wrong. EL has failed as a long-term project for one reason: The project itself was not intended to last. It was merely a form of pressure. Some of the goals were achieved, not all of them, but it was worth the cost. (Enyedy and Tkacz 2011, 117)

While it is clearly arguable as to whether or not the fork was always only a temporary technique, it seems that in hindsight Enyedy is also of the opinion that the greatest impact of the fork was not the creation of the EL.

Over the last two chapters I have considered how forking has been conceived as a political concept and how this conception differs from its historical and contemporary counterparts. I have stressed the contingent nature of executing a fork and made visible the work (as in *working model*) and "sourcery" upon which successful forks rely. I have also stressed that not only is the entire process of forking contingent, but this contingency is distributed in highly asymmetrical ways across different members within a project: not everyone has the same capacity for building allies and creating working models. Together with the fact that forking becomes increasingly difficult as projects scale, forking must be seen as a most uncertain process and cannot be counted on as a mechanism of legitimization. Let me be clear: if forking is the "test of openness," then forking represents yet another way that openness is politically fraught. A successful fork is never guaranteed, but even worse, even successful forks turn out to be messy, full of trivialities and utterly Machiavellian strategies. In our case, the most successful of these strategies turned out to be the technique of the mouthpiece: a technique that once again has a very ambiguous relationship to centralization. The mouthpiece is an individual whose leadership technique is to appear as though authority is radically decentralized. Compared to the *reality* of forking, the discourse on forking—with its emphasis on "maximizing freedom," "radically decentralizing power" and "leaderless-ness"—is unconvincing to say the least.

As an alternative to the existing literature on forking, which depicts forking as a *way out* of possible controversies, my take has therefore been to approach forking as a *way in*. Through a detailed analysis of the controversy that resulted in the fork, I have provided a counter-reading to the always-already-legitimate accounts of force and governance in open projects. In keeping with my general approach, I have set aside the question of legitimacy—it is not a matter of which forms of leadership are better, which vision of Wikipedia is or would be better, or whether the fork had a positive effect on the project—and have instead set myself the more basic task of understanding how force flows through fork controversies. Beginning with a close reading of Enyedy's reply to Sanger, I developed my account of statements to include positive and negative modalities, and also modal relations (where the modality of a statement is dependent on its relation to other statements). I have described the controversy as including different elements and processes: a conflict of statements; the creation of modalities; and the building of allies and realities. I have also described the different actors in controversies—dissenters, dissent assemblages, and spokespersons—and considered the specific technique of representation employed by spokesperson-dissenters in open projects (the mouthpiece). Along the way, I have tried to account for the differences between the language borrowed from Latour and other ANT-based writings and the singularity of Wikipedia (as an open project). Finally, I have suggested that forking can be understood as a stabilizing mechanism, as a way of loading statements and thus making them durable. A consideration of fork controversies thus makes visible forceful statements—statements that come to define a project—in ways different to those considered in chapter 3, but it also shows *how* they gather in strength. Forking is thus also a process of "frame making," the precise opposite of openness. What is perhaps most interesting about this peculiar mode of exit is what it tells us about what lies inside.

CONCLUSION

The Neoliberal Tinge

We are told we are living in neoliberal times. This fuzzy notion forms the backdrop of the political present—it is our common problem. Neoliberalism has come to function in a purely negative manner, attached (often rather sloppily) to whatever is perceived as bad or wrong by that even fuzzier group, "the left." Just as openness is an a priori good, neoliberalism is an a priori bad. But such an account of neoliberalism, which could draw upon any number of things for support—massive income inequality, the erosion of welfare, rampant privatization, ecological devastation, or the naturalization of competition as a social logic, to name a few—nevertheless often overlooks or, rather, fails to take seriously something very important. It fails to consider neoliberalism as a serious response to questions of social and political organization. To be clear, I'm not suggesting that neoliberalism is to be taken seriously as a desirable option. Rather, it must be taken seriously as an intellectual project.[1] Neoliberalism emerged as a coherent response to what were seen as real political problems, such as the centralization of power in the form of the state and the threat of totalitarianism. Neoliberalism was also and remains able to frame problems in such a way that neoliberal solutions appear viable.[2] The failure to recognize this historical achievement has exacerbated confusion about the relationship between neoliberalism and openness, and by turn neoliberalism and the organization of Wikipedia. It is what makes possible the mistaken idea that openness has emerged (instead of re-emerged) as a

---

1. This is one of the reasons that Foucault's (2008) account of neoliberalism and ordoliberalism is exceptional.

2. For examples of neoliberal responses to the global financial crisis and climate change, see Philip Mirowski's *Never Let a Serious Crisis Go to Waste* (2013).

progressive force within an otherwise repressive neoliberal political milieu. Most important, though, it overlooks the fact that openness and neoliberalism have a shared history.

What is this shared history? Neoliberalism was (and remains) one response, one attempted solution, to the problem of the "closed society." As noted, the challenge in the period when Popper and Hayek were writing was the centralization of authority underpinned by (truth) claims about what constitutes the good society. Popper's point of attack was at the level of truth, that is, of epistemology. For Popper, claims to know the truth of what is best for all and for all time results in closed, totalitarian rule. For his part, Hayek proposed the market as a concrete model of organization that is able to overcome this problem of the centralization of authority. If Popper was interested in discovering and refuting what he saw as the philosophical foundations of totalitarianism, Hayek was interested in what it was about market capitalism that made it superior to the alternatives of the day. Thus, the Popperian problem of closure (e.g., centralization, totalitarianism) sees its positive, Hayekian, expression in the validation of the market form. Markets are more than totalitarian-avoiding mechanisms; they are also sites for the active production of freedom.[3] To borrow a term from Foucault (2008, 69), markets are the ultimate "liberogenic" devices and, thus, since roughly the postwar period the ideal market has increasingly become the dominant model for organizing all manner of things.

In a concluding chapter on the history of neoliberal thought, Philip Mirowski has argued similarly that neoliberalism needs to be understood primarily as an epistemology, that is, as involving a set of "epistemic commitments" (2009, 417). The most important of these is the belief that their "vision of the good society . . . will not come about naturally" and rather needs to be "constructed" (434).[4] Of course, and as I have also suggested, this means actively creating markets and market like dynamics wherever political and organizational problems arise. Interestingly, Mirowski (who is a politically charged economic historian) also turns to Wikipedia in these concluding pages, presumably to show how the epistemic commitments of neoliberalism are making their way into the most unlikely of places. He rightly notes, for example, how Jimmy Wales claims to have arrived at the idea for Wikipedia from reading Hayek's "The Use of Knowledge in Society" and also how Wales's view of knowledge in Wikipedia is decidedly Hayekian:

---

3. Of course this coupling of freedom and markets is as old as liberalism itself.

4. Here Mirowski follows Foucault's suggestion that the commitment to actively creating market dynamics is what distinguishes neoliberalism from liberalism.

> Wales subscribes to the precept that objective knowledge is a state rarely attained by any individual because his or her experience is subjective and idiosyncratic; that no individual is capable of understanding social processes as a whole; and that individual beliefs are frequently wonky beyond repair, but given appropriate (market-like) aggregation mechanisms for information, the system ends up arriving at the truth through "free" entry and exit. (2009, 423)

What I want to suggest, however, is something a little more specific. Wikipedia does not emerge primarily from a set of specifically neoliberal "epistemic commitments," but rather from the same "epistemic problem" (of closure) for which neoliberalism is one (flawed) solution. Neoliberalism is one response to the closed society, Wikipedia is another. Put differently, neoliberalism is one articulation of openness, Wikipedia another.

Furthermore, while Wikipedia shares the same "epistemic problem" of neoliberalism—a problem that for Wikipedia now includes the actual consequences of neoliberalism—it nevertheless also draws liberally on eighty-odd years of neoliberal techniques and epistemologies, albeit in a somewhat piecemeal and often less-than-conscious manner. This is why, for example, "collaboration" ends up resembling something like competition. Collaboration is an attempt to mobilize the optimal outcome of competition with none of the negative side effects. Collaboration imagines a market with no losers; it is competition perfected at the level of the distribution of outcomes. Likewise, we saw that the notion of ad-hocracy is drawn directly from 1970s managerial literature, and emerged as a critique of the centralized, hierarchical form of bureaucracy. While forking does not feature specifically in the neoliberal epistemology, we saw how it was described both as *a freedom* (the core freedom in open projects) and as *freedom producing* (as "maximizing the freedom of individuals"). In this sense, forking is a "liberogenic" device par excellence. We also saw how the actual process of forking resembles the creation of a market in the basic sense of generating the conditions of competition between projects. In the final chapter on fork controversy, we saw how strategies of leadership were now orientated around decentralization. More specifically, we saw how it was only possible to speak with force if one could first demonstrate that this force did not originate from the individual. The figure of the "mouthpiece" does not speak on the basis of his or her own and therefore centralized authority; he or she simply reports the outcome of many voices in play. Wikipedia is therefore not identical to neoliberalism, but it does have a certain neoliberal tinge to it. The two share the same epistemic problem of closure and part of Wikipedia's response to this shared problem involves reshuffling the neoliberal deck and mixing in some (more) computational metaphors. Wikipedia is both critique and extension of neoliberalism at the same time.

Zooming out a little, this take on Wikipedia's relationship to neoliberalism (via openness) extends upon existing considerations of the relationship between open projects and neoliberalism, and computers and neoliberalism more generally. Matteo Pasquinelli (2008) has shown, for example, how open projects are *complicit* with new modes of capitalist enterprise (e.g., rent-based models in the form of hardware and ISPs; data aggregation and profiling; individualized advertising). That is, he has shown how open projects, which do not directly produce commodities, have nevertheless been subsumed and "parasited" in such a way that empties out any critical potential. David Golumbia (2009) has rightly pointed out that computers more generally are *incorporated* into forms of neoliberal government and can equally be a force for the centralization of authority. To this general argument we could easily add the fact that neoliberal governments are also readily adopting the language of openness. As we saw in the introduction, government is today imagined as an "open platform" that is itself based on the notion of the bazaar. More recently we have seen how governments are making all kinds of datasets public and query-able, so that "innovators" can create "new apps" for the benefit of society and economy alike.

My own take on neoliberalism, however, aims to extend upon the work of Wendy Chun, by linking "computers to governmentality neither at the level of content nor in terms of the many governmental projects that they have enabled, but rather at the level of their architecture and their instrumentality" (Chun 2011, 9). But instead of considering the relation between computers and neoliberal governmentality in general, as Chun does, I am suggesting something quite specific. Contemporary software and web-based content projects, projects like Wikipedia, face the same political problem as that faced by the neoliberals. They contain within them novel responses to familiar political problems, problems that are fundamentally also about the nature of organization. This means that software, discussion forums, bots, article entries, policies, and guidelines, and everything else that constitutes an open project, must be addressed in relation to the history of this political thought. That is, they must be approached in relation to the problem of organization. It also means that the limits of openness are carried along, from project to project, as long as politics is conceived as the problem of closure.

What I have suggested throughout this book, however, is that this political framework is flawed from the start. The politics of open and closed imagines a world full of binary opposites—open/closed, flat/hierarchical, decentralized/centralized, ad-hocracy/bureaucracy, democratic/totalitarian, and so on—that actively hinders our ability to make sense of the political. The seeds of closure are always already present within the open, but the language of openness doesn't allow us to gain any traction on that closure. Along with critiquing this language of openness, however, my task was also to bring forth the political conditions of openness (via Wikipedia) without buying in to this binary worldview—even in its negative formulation. Identifying the "seeds of closure" isn't enough because it remains in the same binary worldview. The problem of openness isn't that it isn't open; it is that it conceives the world in terms of this question. My task therefore wasn't to show that Wikipedia is actually closed, hierarchical, centralized, bureaucratic, or totalitarian, but rather to try to think politically differently. How can one theorize the conditions of working together while refusing the choice between "managerial commands" and "market signals," or when neither "spontaneity" nor "hierarchy" reflects what's going on? How can one describe organizational governance without recourse to ad-hocracy or bureaucracy? What can we learn from the exit strategy of forking if we don't automatically accept its function as a "safety net" and thus as a mechanism of legitimization? That is, if we don't believe in its capacity to "maximize freedom," "de-monopolize power," and "defend against tyranny"?

In the shadow of that "spontaneous," "stigmergic," "unmanaged" and "self-organized" mode of working together, collaboration, we find practices of framing; we find damages, wrongs, the possible outcomes of differends and boundary objects, and a whole repertoire of "statement games." We find the sorting of statements and people. In the shadow of that theory of organization where the very question of organization is obscured, ad-hocracy, we find forceful statements that flow throughout an entire formation; through its "files," "scribes" and "material apparatuses" alike. We find statements whose history extends the life of Wikipedia (NPOV) and we see how statements become "loaded" in different ways in order to gather force. We discover a method for tracing the singularity of organizational-governmental forms, *ex corpore*. And in the shadow of that "freedom maximizing," "de-monopolizing," "tyranny defending" theory of political exit, forking, we find radical contingency. We find acts of "sourcery," "working models of equivalence" based on a shared "computational" worldview, and the problem of scale. We find the messiness of fork controversies, more statement games, negative modalities, dissenters, and new techniques of leadership (spokespersons and mouthpieces). We discover that forking is equally a practice of making projects and their most forceful statements more durable; it is a practice of shoring things up. In other words, what we find in Wikipedia is a language from which to speak back to openness.

There are now innumerable iterations of openness, a thousand different politics that differ as much as they resonate with one another. What are the micro-politics of these projects? Which principles define their frames? What are their mechanisms of legitimization? What are their techniques of leadership and representation? These questions lead away from openness, into the politics of organization.

Appendix A: Archival Statements from the Depictions of Muhammad Debate

Statements in Favor of Inclusion

1. Censorship
    1.1. "Wikipedia is not censored." (http://en.wikipedia.org/wiki/Talk:Muhammad)
    1.2. "This is not censure-anything-that-someone-could-find-offensive-pedia, and it doesn't matter here what Islam forbids or not." (Archive_2)
    1.3. "Just as "Wikipedia is not censored for minors," "Wikipedia is not censored for iconoclasts," either." (Archive_2)
    1.4. "I'm sorry, but it is absolutely unacceptable to censor Wikipedia just because something is offensive to a religious group." (Archive_4)
    1.5. "I am tired of seeing WP censored because someone thinks a image of the Quran with a woman next to it is offensive or a painting of some prophet as blasphemy. There are plenty of images on WP that I think we could do with out . . . So my bottomline is this. . If you think an image is offensive, DON'T LOOK AT IT!! If you feel compelled to look at it and don't want to see it, TURN OFF YOUR COMPUTER! Enough of this people. Grow the hell up." (Archive_4)
    1.6. "Wikipedia does not censor itself for a creed, religion, or cause. Leaving a picture out of a biography of a person, when there are many avaliable, does an injustice to the article. Wikipedia is not run by Muslims, nor does it have to follow the wishes of Muslims. Wikipedia has many things on it critical of Islam, shall we censor them too, so as to not anger any Muslims?" (Archive_4)
    1.7. "When respect and understanding means suppressing information, then they amount to censorship, and I'll fight censorship to the death." (Archive_4)

1.8. "WP is a secular project—that Mohammed's depiction keeps creeping down before it gets deleted over and over again is shameful censorship." (Archive_10)

2. Respect
   2.1. "Your beliefs are not mine, and you have no right to tell me what I can or can not see, say, hear or do. You ask us to respect your principles, but I have to ask: why will you not respect mine?" (http://en.wikipedia.org/wiki/Talk:Muhammad)

3. Not religious
   3.1. "The Islam's (or the Muslims') stance on whether pictures of Muhammad are acceptable or not is utterly irrelevant—WP is not a religious project, so religious rules cannot apply. Otherwise, anything that could be considered non-compliant with the rules of Islam would have to be deleted—any criticism of Muhammad, any doubts about whether he had a vision / an encounter with Gabriel / insert random religious belief here." (Archive_2)
   3.2. "There are many pictures, both contemporary and historical, of Mohammed out there, I do not see why, as Wikipedia is a encyclopaedia that is supposed to be secular and impartial, why we should bow to the wishes of Muslims simply because it is 'offensive' to them. No one can be as offensive to Muslims as the Quran is to non-muslims, ergo, asking not to have a picture of Mohammed (offensive, hate-filled language) is tantamount to me saying that they should remove verses from their Quran that offend me." (Archive_3)
   3.3. "If we are going to start scrubbing articles on religions to make sure that nothing in the offends adherents of that religion, we are going to end up with some pretty short articles." (Archive_3)
   3.4 "... pictures of Muhammed are only disturbing people entertaining a certain belief system. To the rest those pictures are about as offensive as a picture of an orange. Anonymous is right, to people of many belief system the theory of evolution is extremely offensive. What makes us keep that article and scrap the pictures of Muhammed? The answer is scary." (Archive_4)
   3.5. "This is an encyclopedia, not the Quran. There is no reason not to have a picture." (Archive_4)
   3.6. "Many of the editors wanting removal of all pictures (or just this one) are advocating outright censorship due to their religious be-

liefs. This is not in keeping with the policy or spirit of Wikipedia." (Archive_8)

3.7. "The picture that has been the subject of much reverting is neither obscene or profane. Nor is it defamatory. That leaves the question of whether it is offensive. While this may seem subjective, it is really is not. The only way this picture is offensive is from a religious standpoint. Which is making a lot of liars out of those people who delete this picture for this reason but yet lie in their edit summary." (Archive_8)

3.8. "Even if every Muslim on earth objected to us having a picture here, it should still cut no ice. This is the English language Wikipedia, and every single English-language speaking country on earth has an ancient and unconditional tradition of free speech. This means that we should be able to include whatever picture we like to depict the written content, whatever the consequences." (Archive_8)

3.9. "Dear Muslim Wikipedians, you ought to take note of the fact that Wikipedia is a non-Islamic encyclopedia. That you happen to dislike portraits of Muhammad has absolutely no bearing on what is allowed here. You simply have no right to demand that no painted portraits of Muhammad be shown here! (Anyway, the claim that any portraits of the Prophet have always been universally forbidden in Islamic history is a lie. There are many such depictions made by Muslims.)" (Archive_10)

4. Not all Muslims are against the images
    4.1. "It was even created by a Muslim." (Archive_2)
    4.2. "It is Muslim iconography! Islam is not monolithic, and if some Muslims are offended by other Muslims views, that is hardly Wikipedia's fault, we are simply here to document all of them, regardless of whether some are offensive to some."(Archive_2)

5. Relevance
    5.1. "It's a beautiful public domain painting of the person that this article discuss, so it is entirely relevant and there is no excuse to censure it." (Archive_2)
    5.2. "If we had Muhammad's mugshot, of course it would be relevant and I would support adding it. We don't. All we do have are artists' impressions. If we are to add those, they have to be notable, wrt Muhammad. The Persian Miraj image is relevant. If you like we can add

other images from the Siyer-i Nebi. If you have other notable Muslim artwork, suggest it." (Archive_3)

5.3. "The discussion about a picture Mohamed is totally misses the point that we are writing articles for an Ecyclopedia." (Archive_4)

5.4. "Including an image of Muhammad in an encyclopedic entry on Muhammad seems completely natural to me. The argument from religion shouldn't be of relevance for the discussion, in my opinion." (Archive_4)

5.5. "To start a discussion concerning the image itself:
   1. It depicts an important historical event
   2. It is penned by someone from the very culture of the event—giving it additional authenticity
   3. It is part of a truly masterful historical compendium—while all those self-shot flower photos and DIY-SVG-coats-of-arms are well done, Wikipedia can only benefit from featuring some examples of superbly crafted medieval art from centuries ago." (Archive_8)

5.6. "Pictures have value because they give us a sense of historicity, even if the picture may be inaccurate. It tells us what people of another era were thinking about historical figures. Which is very valuable in itself. It helps our mind by giving us 'context' for the surrounding text. It gives us a perspective that this person was even thought about enough that they took the trouble to make a picture of him, and the setting that they imagined Muhammad in." (Archive_8)

5.7. "Its customary for any encyclopedia known to me (WP, Britannica, Larousse, Brockhaus) to illustrate the subject if possible. If so, the illustration is right on the entries top." (Archive_10)

6. Precedent
   6.1. "That stuff is even in the Persian Wikipedia: [7] so we should have no problem with it here." (Archive_4)
   6.2. "As a side note, there's are images of 'God' in the God entry and images of Jesus in the Jesus entry, although the Hebrew/Christian Bible also include a prohibition of images of God. Although the prohibitions in bible as well as those in the Koran are ambiguous." (Archive_4)

7. NPOV
   7.1. "You misunderstand me. Your comments are about why some Muslims believe that images of Muhammad are undesirable, but they've

no obvious application to wikipedia because wp is NPOV." (Archive_5)

7.2. "Wikipedia is an NPOV encyclopedia, and therefore aims to present an unbiased presentation of all notable viewpoints. This includes secular viewpoints, and it certainly includes images which are notable works of art, even if they may be offensive to some readers." (Archive_10)

8. Barbecue analogy

8.1. "The image is a valuable contribution to the whole article—I made that case above and nobody bothered objecting my reasoning. Let me specify:

"Example 1: You're having a perfectly legal barbecue in your garden, but your neighbour is bothered by the noise & smell. He complains to you, but as a reaction, you turn up the volume & nudge your grill even further towards his garden. The case is clear—someone doing this is a rude and annoying jerk! But did this happen here? No, because:

"Example 2: Point of departure as above, but this times your neighbour complains about barbecuing steaks & burgers as such because he is a hardcore vegetarian. He considers killing animals murder & eating them the devouring of rotten corpses. His stomach turns on seeing you flipping your burgers and consuming them. Addressing his complaints you tell him "Sorry dude, but I'll keep on doing what I'm doing"—so, is the barbecue guy a rude jerk just like the guy from Ex. 1? I don't think so. He has no obligation to quit grilling, his actions are both legal and justified, i.e. in accordance with generally accepted social norms. It might be considered polite to stop your barbecue because your neighbour already feels nauseous—but get real, the guy obliged to give in is the neighbour, not the barbecue guy!

"And this is exactly the case! It's not some nasty actions, which are leading to frictions here, but the individual perception of an otherwise perfectly acceptable behaviour. So, this is, in a nutshell, why I object to deletion of the picture and strongly reject grounds like 'It's rude' or 'You do this just to bother me.'" (Archive_8)

9. Consensus

9.1. "Well, it's not a consensus as in 'everybody agrees,' but one like in 'no valid objections were raised beyond a general averseness to pictures.'

I made several points, none of them were addressed as regards content, only with variants of 'I don't like it.' Since the image in question lifts the article's quality both in content and form, I suggest to reinsert it and request FayssalF to unprotect the article in order to do so." (Archive_8)

Statements against Inclusion

10. Offensive to Muhammad
    10.1. "Please take that picture off, because it is an offence to him." (http://en.wikipedia.org/wiki/Talk:Muhammad)

11. Images are offensive to Muslims
    11.1. "Pictures have a condemnable place in our religion" (http://en.wikipedia.org/wiki/Talk:Muhammad)
    11.2. "No, this is not 'is not censure-anything-that-someone-could-find-offensive-pedia,' but neither is it 'Controversial-for-the-sake-of-it-pedia.'" (Archive_2)
    11.3. "We have two options. Include the picture, or don't. If we include the picture, we get a painting, and we offend various people. If we don't, we don't get a painting, and we don't offend various people. I see no significant advanatage to offending people, nor to having the painting." (Archive_2)
    11.4. "Is there a firm wikipedia policy on inserting religiously offensive material solely for the reason of annoying others? If so, could someone point me to it so I could bring up the point at a Mormon article where people do the same thing?" (Archive_4)
    11.5. "The proposal to put a picture into the article has come up several times, with the end result being that a picture is not necessary. Because the portrayal of the Prophet Muhammad (peace be upon him) is blasphemous in Islam and because no portrayal of the Prophet is accurate, there is no reason to put a portrayal of the Prophet into this article except to offend Muslims." (Archive_4)
    11.6. "Given the mayhem and deaths associated with the current backlash against the Danish cartoons (and, more importantly, their provocative reprinting), I think we can safely deduce that such depictions (of created things, but particularly of Muhammad) are seriously offensive to many mainstream Muslims—as offensive as, say, images of paedophilia. Thus, it seems reasonable to decide that the definitive wikipedia article about the prophet of Islam need not include an image of

him, out of respect for the heart-felt beliefs of members of the second largest religion in the world. This is not censorship, it is consensus-building respectfulness." (Archive_4)

11.7. "Are you referring to a picture of Prophet Muhammad? That is because muslims believe that representing Muhammad as an image, in the form of drawing, painting, idol making etc is forbidden, because then it gives the impression that muhammad is an idolatry figure to be worshiped." (Archive_5)

11.8. "Please do not upload any image of Muhamamd, that may be any drawing or any imaginary image. Because Muslims dislike this and in relagion Islam to Portrait any image of Muhamamd (PBUH) is prohibited. So please avoid it. Thank you." (Archive_10)

12. Respect

12.1. "Please respect our religious principles. Blocking those pictures form our browsers will not satisfy us. They'll still be visible to others! . . We don not wany ANYONE to have any picture in their mind as to what our beloved prophet (s.a.w.w) looked like! . ." (http://en.wikipedia.org/wiki/Talk:Muhammad)

12.2. "I have the strong impression that the determination to fill the article with images of Muhammad has less to do with conveying information (that's in a breakout article, on Depictions of Muhammad) or creating a visually pleasant article (calligraphy can do that) than it does with a desire to piss off Muslims. "Nyah nyah nyah you can't censor me!" Seems to me that there's a WP principle about not creating ruckuses to make a point. WP:Point." (Archive_10)

13. Accuracy

13.1. "Fact is that the overwhelming vast majority of Art venerating Muhammad is Calligrapy. If you want an accurate representation of art venerating him, then Calligraphy is the way to go." (Archive_2)

13.2. "I also remain skeptical that an accurate drawing of the Prophet Muhammad (pbuh) due to the fact that he lived fourteen centuries ago . . ." (Archive_3)

13.3. "As far as I know, there are no factual portraits of Muhammad accepted by Muslims." (Archive_3)

13.4. "No picture of the Prophet is accurate." (Archive_4)

13.5. "If there was a well known or regonisable image of Mohammed I would support its inclusion. However due to tradition there isn't, therefore it is not necessary to have one. I think the policy of not disrupting the Wiki to make a point applies here." (Archive_4)

13.6. "If the goal of an encyclopedia is to be accurate then only things that are true should be in it. This is the logic I am pursuing.

"If you look up the word 'inaccuracy' and 'true' on dictionary.com the you will find that inaccuracy is defined as 'containing or characterized by error' and that the word true is defined 'Consistent with fact or reality.'

"Since none of the pictures depicting prophet Mohammed (PBUH) are actually of him, but most probably of some other person (or just made up) then if those images are to be included then one can not suggest that the purpose of this encyclopedia is to be accurate. If there is no desire to make Wikipedia accurate then really it is meaningless and has no value." (Archive_5)

13.7. "Any drawing or picture would never be an accurate physical representation of Muhammad. There are descriptions of him in hadiths. But noone who is alive knows what he looked like." (Archive_5)

14. Civility

14.1. "I removed the painting of Muhammed from the article. I have done this because of my interpretation of WP:Civility. [. . .] That having been said, my understanding of Islam is that paintings and pictures, images, and so on are avoided or prophibited because it is seen as dangerously close to idolaltry. So, I think, once we acknowledge that, to have a picture of Islam's most important figure/prophet/personage would be, if not blasphemous, atleast very impolite. There is a commons link for those who wish to look for these images, and removing it from the article keeps offending persons with the highest intrest in the article at a minimum.

"This isn't about right or wrong, or censorship or anything else. I simply think its a matter of being polite and civil." (Archive_2)

15. Quality/Relevance

15.1. ". . . it doesnt improve the informative quality of article ." (Archive_3)

15.2. ". . . this proposal will not add any value to the article as others have mentioned. The picture of Moses would, in my opinion, be better off on the Michaelangelo article rather than the Moses one. It doesn't add anything to what Moses looked like, but rather it gives insight into the works of Michaelangelo. The pictures on Jesus and Confucius also do not adding anything of worth to those respected articles." (Archive_3)

15.3. "Unless the picture provides a valuable, educational purpose, it should otherwise not be included in the article." (Archive_5)

15.4. "Relevant guideline: 'Words and images that might be considered offensive, profane, or obscene by other Wikipedia readers should be used if and only if their omission would cause the article to be less informative, relevant, or accurate, and no equally suitable alternatives are available. Including information about offensive material is part of Wikipedia's encyclopedic mission; being offensive is not.'
"It is already evident from the above discussion that the inclusion will be considered offensive by a lot of people, so the question becomes whether leaving it out would cause the article to become 'less informative, relevant, or accurate.' While I agree that the image might be a valid embellishment to the article, I don't see it as reducing the informativeness, relevance or accuracy in any way if it were left out." (Archive_9)

15.5. "The image of Muhammad preaching serves absolutely no purpose. Why is it here?" (Archive_10)

16. Consensus

16.1. "I disagree, Hun. It seems to me there is no consensus to insert this image." (Archive_8)

Solution Attempts

17. Recognizing both sides

17.1. "I dunno what I think about replacing the picture. On the one hand, I don't want to be needlessly offensive; on the other hand, I really don't want Wikipedia to censor stuff that only a small minority of viewers would find problematic. It seemed to me that by presenting the picture but also saying the some Muslims found such pictures offensive, we did present both POVs." (Archive_2)

17.2. "Therefore, dear aniconists, instead of complaining of the images we have, do upload us some nice images of notable 'Muhammad' calligraphy, and I will certainly suppport giving those precedence over Shia portraits (although at least a single Persian image should remain here for balance)." (Archive_10)

18. Link

18.1. "however, we could comprimise by having a link to his picture (and not actually having his picture in this article)." (Archive_3)

18.2. "I would like to propose a solution that would minimise offence. Would it not be possible for the main article to have an icon (forgive the religious pun) indicating that a click upon it would lead to an image? There are all sorts of topics that could benefit, not just Muhammed. Many people would prefer to gain information about a distressing topic through words and not images—car crash, for example, or starvation. Others would like to read about human anatomy or diseases without having to look at the evidence—not just genitalia, but internal organs can be considered "private parts"! Some modern art installations can fall into this category too. So if there's a fair chance of an image offending readers, why not include it on a separate page all to itself, just one click away. No censorship; lots of civility and consideration." (Archive_7)

19. Position

19.1. "The concensus reached on the article about Bahá'u'lláh was that the image should be at the bottom of the page where believers could avoid it and that the summary at the begining would state that the image was at the bottom with the dual purpose of 'warming' those who do not want to see the image and informing people who are curious as to why the image does not appear in the top right." (Archive_4)

20. Notice

20.1. "Maybe a sort of Viewer Discretion warning at the top of the page would help?" (Archive_10)

Appendix B: Selections from the Mediation Archives

Statements of Position

For:

> Non-professor Frinkus: "I hope this was not already covered: I believe, if the image is clearly relevant to the article, it should be included. If people find depictions offensive for any reason (regardless of size of that group), that should not be a consideration beyond ensuring that the image has some historical/informative value before being used. For example, if people do not like images of a particular person, one should not source a recently produced generic image just for the sake of including an image; however, images that adds something relevant to the article is fine regardless of other objections. This would apply to all controversial images." Nonprof. Frinkus 20:30, 4 November 2006 (UTC) (Mediation _Archive_2)

> Chowbok: "Despite nobody knowing what they actually looked like, the articles on Homer, Jesus, Alexander the Great, Zoroaster, Socrates, William the Conquerer, Moses, and Genghis Khan all have images at the top of the page. Clearly, this means that adding an image to the top of the Mohammad article wouldn't be done simply to antagonize Muslims; it would just be in keeping with standard Wikipedia practice. On the other hand, were we to not include an image on the Muhammad page, the only reason we would be doing it would be to capitulate to a vocal minority. We don't let people remove appropriately-included profanity or nudity from articles, why should we let people remove an unquestionably appropriate image? If we agree to leave the image out in this case, or even to push it to the bottom of the page, we might as well completely throw out

WP:NOT#Wikipedia is not censored."—Chowbok 21:29, 4 November 2006 (UTC) (Mediation_Archive_2)

Against:

Irishpunktom: "Ok, firstly, there is no breach in wikipedia this is not a policy issue. Considering Images are offensive to a substantial amount of people (and I have seen no-one dispute that, so i presume it is agreed all round?), surely we need a good reason to include any images into the biography article? . Knowing that such images of Muhammad are offensive has been used as a tool to attack Muslims with; Images that no-one knows a thing about are being added with no context, false attribution, and an ignorance of the artist; added solely to offend; Wikipedia should not gratutitiously offend. The majority of art surrounding other religious figures is art created by artists who adhere to the religion associated ith that person, and the majority of Islamic art concerning Muhammad is calligraphy, of which there is a shortage in this biography."—Irishpunktom \talk 00:10, 5 November 2006 (UTC) (Mediation_Archive_2)

IbrahimFaisal: "I do not support any picture of Muhammad in the Muhammad article because: no picture exists that has resemblance or related to Muhammad. All early surviving biographies were written by Muslims and they do not have any pictures of Muhammad. The oldest picture we found so far is of 13th century that is 7/6 hundred years after Muhammad death. Yes these might be historical but how could be they related with Muhammad biography directly? ALL Sunni Muslim (90% of total Muslims) do not like Muhammad picture of any kind and many Shia too (if not all). There are alternative, for example pictures not showing Muhammad but only events and calligraphy pictures then why to offend other people? By including Muhammad picture we discourage Muslims to contribute in Muhammad article as well as in wikipedia and we are NOT improving the article quality too by some imaginary picture."—اباربيم 16:38, 5 November 2006 (UTC) (Mediation_Archive_2)

BhaiSaab: "I generally agree with what Ibrahim has stated. I'd also like to note that even though 'Wikipedia, and for that matter, other encyclopedias, have good precedence for showing the pictures of ancient historical figures,' I have not seen pictures of the Prophet in the articles on him in Encarta, World Book Encyclopedia, Columbia Encyclopedia, or Britannica." BhaiSaab talk 22:40, 5 November 2006 (UTC) (Mediation _Archive_2)

TruthSpreader: "I believe that the picture under discussion does not have any informational value but it actually gives wrong information. According to hadith literature, Muhammad used to sit with people and it is also reported that when someone from outside used to come, it was difficult for him to find the prophet because he used to mix with others in a cordial fashion. What the author of the picture is implying, just simply doesn't pass NPOV." TruthSpreaderTalk 11:37, 6 November 2006 (UTC) (Mediation _Archive_2)

Mediation Summaries of Positions

First attempt:

Clarity

"Okay, I have thought a lot about how to best approach this case, because there are many issues at play here. All of them are subjective, and all of them have been debated ad infinitum. Therefore, I don't see much value in debating them again here, because in my experience, everyone brings out the same arguments and nothing is agreed upon. So, here's what I propose.

"I am going to summarize what I think are the major sides of the issue:

"Encyclopedic depictions of Muhammad should be included in the article. Removal on the basis of relevance or notability may be discussed on a per-image basis.

"Depictions of Muhammad should not be included in the article since they are offensive to many Muslims who read Wikipedia, and the depictions may be made available in a separate article (such as Depictions of Muhammad).

"Now. Sign below to indicate whether you agree or disagree that these two points adequately summarize the debate. If everyone agrees, we can proceed to compromise." (Mediation_Archive_3)

Second attempt:

Refining positions

"Okay, based on all of the discussion in the previous section and my interactions with BostonMA, I am going to try to refine the summary of positions below:

"Encyclopedic depictions of Muhammad should be included in the article, and held to defined standards of notability and relevancy. Standards will be defined in this mediation.

"Depictions of Muhammad are not informative (and by extension, not encyclopedic) because the physical appearance of Muhammad is unknown, and the depictions are offensive to many Muslims. As such, the depictions should not appear in the article.

"I am asking that you agree that you fit into one of these categories. It is not necessary to state agree or disagree unless your position has changed. I am simply trying to satisfy everyone so we can move on."—Aguerriero (talk) 15:55, 9 November 2006 (UTC) (Mediation_Archive_3)

Statement of Criteria for Inclusion

Image Criteria: Sandbox

"Okay everyone, since no one seems to object to BostonMA's edit of one of the position statements, I think we can move on. I considered for quite some time whether it would be more useful at this time to discuss the criteria for including depictions of Muhammad in this article, or to discuss where to include them or whether to include them at all.

There are two editors who, in their statements, indicated that depictions might belong elsewhere (Striver) or do not belong at all (IbrahimFaisal/ALM scientist and BhaiSaab). However, I think it will be more useful to get the discussion of criteria out of the way first, and then proceed to the other discussion since editors who are currently opposed to any kind of inclusion might reconsider if satisfactory criteria are developed? Does that makes sense? I hope so . . . it's Friday afternoon and my brain power is waning.

That being said, I am going to propose criteria. If you agree, awesome, sign. If you don't agree, create a subheading with your name and your proposed changes, or 'this is all bunk,' whatever your position is. I will be largely absent over the weekend, so I will check in Monday and view everyone's comments."—Aguerriero (talk) 21:39, 10 November 2006 (UTC)

Proposed criteria for including depictions of Muhammad in the Muhammad article:

"The image is properly attributed with its title, creator, and origin (museum, manuscript, etc) using a reliable and neutral source (see WP:RS). If any of those are unknown, the citation should acknowledge that it is not known.

"The beings and/or events depicted in the image are properly described using a reliable source: a citation explaining what it depicts (see WP:OR).

"The image's notability is asserted using a reliable source: a citation explaining why the image is notable in reference to Muhammad (see WP:N) (removed)

"The use of the image conforms to WP:NPOV.

"The use of the image meets the standards of WP:Profanity."

—Aguerriero (talk) 21:39, 10 November 2006 (UTC) (Mediation_Archive_4)

References

Adorno, T., H. Albert, R. Dahrendorf, J. Habermas, H. Pilot, and K. Popper. 1976. The Positivist Dispute in German Sociology. Translated by G. Adey and D. Frisby. London: Heinemann Educational Publishers.

Akrich, M., M. Callon, and B. Latour. 2002. "The Key to Success in Innovation Part II: The Art of Choosing Good Spokespersons." *International Journal of Innovation Management* 6 (2): 207–25.

Armstrong, C. 2010. "Emergent Democracy." In *Open Government: Collaboration, Transparency, and Participation in Practice*, ed. D. Lathrop and L. Ruma, 167–76. Cambridge: O'Reilly Media.

Bateson, G. 1972. *Steps to an Ecology of Mind: Collected Essays in Anthropology, Psychiatry, Evolution, and Epistemology*. London: Intertext Books.

Bauwens, M. 2005a. "P2P and Human Evolution: Peer to Peer as the Premise of a New Mode of Civilization." Retrieved March 14, 2010, from http://www.altruists.org/f870.

———. 2005b. "The Political Economy of Peer Production." *CTHEORY*. http://www.ctheory.net/articles.aspx?id=499.

BBC News. 2006. "Riots in Nigeria Leave Many Dead." *BBC News*, accessed July 5, 2011, from http://news.bbc.co.uk/2/hi/africa/4738726.stm.

Becker, B. 2010. "A Revised Contract for America, Minus 'With' and Newt." *New York Times*, April 15, A19, http://www.nytimes.com/2010/04/15/us/politics/15contract.html?_r=2.

Benkler, Y. 2006. *The Wealth of Networks: How Social Production Transforms Markets and Freedom*. New Haven, CT: Yale University Press.

Bennis, W. 1965. "Beyond Bureaucracy: Will Organization Men Fit the New Organizations?" *Society* 2 (5): 31–35.

Bergson, H. 1935. *The Two Sources of Morality and Religion*. Translated by R. A. Audra, C. Brereton, and W. H. Carter. New York: H. Holt and Company.

Berry, D. M. 2008. *Copy, Rip, Burn: The Politics of Copyleft and Open Source*. London: Pluto.

Beschastnikh, I., T. Kriplean, and D. W. McDonald. 2008. "Wikipedian Self-governance in Action: Motivating the Policy Lens." Paper presented at the ICWSM 2008, Seattle, WA. http://www.cs.washington.edu/homes/ivan/papers/icwsm08.pdf.

Boltanski, L., and L. Thévenot. 1999. "The Sociology of Critical Capacity." *European Journal of Social Theory* 2 (3): 359–77.

———. 2006. *On Justification: Economies of Worth*. Translated by C. Porter. Princeton, NJ: Princeton University Press.

Bowker, G. C., and S. L. Star. 1999. *Sorting Things Out: Classification and Its Consequences*. Cambridge, MA: MIT Press.

Boyle, J. 2008. *The Public Domain: Enclosing the Commons of the Mind*. New Haven, CT: Yale University Press.

Bruns, A. 2008a. *Blogs, Wikipedia, Second Life, and Beyond: From Production to Produsage*. New York: Peter Lang.

———. 2008b. "The Future Is User-Led: The Path towards Widespread Produsage." *Fibreculture Journal* 11. http://eleven.fibreculturejournal.org/fcj-066-the-future-is-user-led-the-path-towards-widespread-produsage/.

Butler, B., E. Joyce, and J. Pike. 2008. "Don't Look Now, But We've Created a Bureaucracy: The Nature and Roles of Policies and Rules in Wikipedia." Paper presented at the 26th annual SIGCHI Conference on Human Factors in Computing Systems. Florence, Italy.

Butler, J. 2009. *Frames of War: When Is Life Grievable?* London: Verso.

Callon, M. 1986. "Some Elements of a Sociology of Translation: Domestication of the Scallops and the Fishermen of St Brieuc Bay." In *Power, Action and Belief: a New Sociology of Knowledge*, ed. J. Law, 196–223. London: Routledge.

———. 1998. *The Laws of the Markets*. Oxford: Blackwell Publishers/The Sociological Review.

Capocci, A., V. D. P. Servedio, F. Colaiori, L. S. Buriol, D. Donato, and S. Leonardi. 2006. "Preferential Attachment in the Growth of Social Networks: The Internet Encyclopedia Wikipedia." *Physical Review E* 74 (3). http://arxiv.org/pdf/physics/0602026.pdf.

Carr, N. 2011. "Questioning Wikipedia." In *Critical Point of View: A Wikipedia Reader*, ed. G. Lovink and N. Tkacz, 191–202. Amsterdam: Institute of Network Cultures.

Chen, S. 2011. "The Wikimedia Foundation and the Self-governing Wikipedia Community." In *Critical Point of View: A Wikipedia Reader*, ed. G. Lovink and N. Tkacz, 351–71. Amsterdam: Institute of Network Cultures.

Chun, W. H. K. 2008. "On 'Sourcery,' or Code as Fetish." *Configurations* 16:299–324.

———. 2011. *Programmed Visions: Software and Memory*. Cambridge, MA: MIT Press.

Clark, D. C. 1992. "A Cloudy Crystal Ball—Visions of the Future." Paper presented at the Proceedings of the 24th Internet Engineering Task Force, Massachusetts Institute of Technology, Cambridge, MA.

Cohen, N. 2011. "Define Gender Gap? Look Up Wikipedia's Contributor List." *New York Times*. Retrieved from http://www.nytimes.com/2011/01/31/business/media/31link.html.

Coleman, G. 2004. "The Political Agnosticism of Free and Open Source Software and the Inadvertent Politics of Contrast." *Anthropology Quarterly* 77 (3): 507–19.

Cornforth, M. 1968. *The Open Philosophy and The Open Society; A Reply to Dr. Karl Popper's Refutations of Marxism*. New York: International Publishers.

Cramer, F. 2011. "A Brechtian Media Design: Annemieke van der Hoek's Epicpedia." In *Critical Point of View: A Wikipedia Reader*, ed. G. Lovink and N. Tkacz, 221–25. Amsterdam: Institute of Network Cultures.

Darnton, R. 1987. *The Business of Enlightenment: A Publishing History of the Encyclopédie, 1775–1800*. Cambridge, MA: Belknap Press of Harvard University Press.

de Certeau, M. 1984. *The Practice of Everyday Life*. Translated by S. Rendall. Berkeley: University of California Press.

De Cock, C., and S. Bohm. 2007. "Liberalist Fantasies: Zizek and the Impossibility of the Open Society." *Organization* 14 (6): 815–36.

Debord, G. 1994. *The Society of the Spectacle*. Translated by D. Nicholson-Smith. New York: Zone Books.

Deleuze, G., and F. Guattari. 1988. *A Thousand Plateaus: Capitalism and Schizophrenia*. Translated by B. Massumi. London: Continuum.

Derrida, J. 1997. *Of Grammatology*. Translated by G. C. Spivak. Baltimore: Johns Hopkins University Press.

Descy, D. E. 2006. "The Wiki: True Web Democracy." *TechTrends* 50 (1): 4–5.

Deutschman, A. 2007. "Why Is This Man Smiling?" *Fast Company*. Retrieved from http://www.fastcompany.com/magazine/114/features-why-is-this-man-smiling.html?page=0%2C2.

Eco, U. 1989. *The Open Work*. Translated by A. Cancogni. Cambridge, MA: Harvard University Press.

Elliott, M. 2006. "Stigmergic Collaboration: The Evolution of Group Work." *MC Journal: A Journal of Media and Culture* 9 (2). http://www.inf.ucv.cl/~bcrawford/Cuesta_Olivares/Stigmergy/stigmergy%20work.pdf.

———. 2007. "Stigmergic Collaboration: A Theoretical Framework for Mass Collaboration." PhD diss., University of Melbourne, Melbourne. Retrieved from http://dlc.dlib.indiana.edu/dlc/bitstream/handle/10535/3574/elliott_phd_pub_08.10.07.pdf.txt?sequence=2.

Enyedy, E., and N. Tkacz. 2011. "'Good Luck with Your WikiPAIDia': Reflections on the Spanish Fork of Wikipedia." In *Critical Point of View: A Wikipedia Reader*, ed. G. Lovink and N. Tkacz, 110–18. Amsterdam: Institute of Network Cultures.

Ernst, W. 2012. *Digital Memory and the Archive*. Minneapolis: University of Minnesota Press.

Filipacchi, A. 2013. "Wikipedia's Sexism Towards Female Novelists." *New York Times*. Retrieved from http://www.nytimes.com/2013/04/28/opinion/sunday/wikipedias-sexism-toward-female-novelists.html?_r=1and.

Fogel, K. 2005. *Producing Open Source Software: How to Run a Successful Free Software Project*. Sebastopol, CA: O'Reilly Media.

Forte, A., and A. Bruckman. 2008. "Scaling Consensus: Increasing Decentralization in Wikipedia Governance." Paper presented at the HICCS, IEEE Computer Society, Washington, DC.

Foucault, M. 1972. *The Archaeology of Knowledge*. Translated by A. M. S. Smith. London: Tavistock Publications.

———. 1977. *Discipline and Punish: The Birth of the Prison*. Translated by A. Sheridan. New York: Pantheon Books.

———. 1980. *Power/Knowledge: Selected Interviews and Other Writings, 1972–1977*. Translated by C. Gordon, L. Marshall, J. Mepham, and K. Soper. New York: Pantheon Books.

———. 1996. "What Is Critique?" In *What Is Enlightenment?*, ed. J. Schmidt, 382–98. Berkeley: University of California Press.

———. 2008. *The Birth of Biopolitics: Lectures at the College de France, 1978–1979*. Translated by G. Burchell. New York: Picador (Palgrave Macmillan).

Fuster Morell, M. 2011. "The Wikimedia Foundation and the Governance of Wikipedia's Infrastructure." In *Critical Point of View: A Wikipedia Reader*, ed. G. Lovink and N. Tkacz, 325–41. Amsterdam: Institute of Network Cultures.

Galloway, A. R. 2004. *Protocol: How Control Exists After Decentralization.* Cambridge, MA: MIT Press.

Geiger, R. S. 2009. "The Social Roles of Bots and Assisted Editing Tools." Paper presented at the 2009 International Symposium on Wikis and Open Collaboration, Orlando, FL.

———. 2010. "The Work of Sustaining Order in Wikipedia: The Banning of a Vandal." Paper presented at the 2010 Conference on Computer Supported Cooperative Work, Savannah, GA.

———. 2011. "The Lives of Bots." In *Critical Point of View: A Wikipedia Reader*, ed. G. Lovink and N. Tkacz, 78–93. Amsterdam: Institute of Network Cultures.

Goffey, A. 2008. "Algorithm." In *Software Studies: A Lexicon*, ed. M. Fuller, 15–20). Cambridge, MA: MIT Press.

Goffman, E. 1974. *Frame Analysis: An Essay on the Organization of Experience.* Cambridge, MA: Harvard University Press.

Golumbia, D. 2009. *The Cultural Logic of Computation.* Cambridge, MA: Harvard University Press.

Hancock, T. 2010. "OpenOffice.org Is Dead, Long Live LibreOffice—or, The Freedom to Fork." *Free Software Magazine.* Retrieved from http://www.freesoftwaremagazine.com/columns/openoffice_org_dead_long_live_libreoffice.

Hardt, M., and A. Negri. 2004. *Multitude: War and Democracy in the Age of Empire.* New York: Penguin Press.

Hartman, M. 1986. "Hobbes's Concept of Political Revolution." *Journal of the History of Ideas* 47 (3): 487–95.

Harvey, D. 2005. *A Brief History of Neoliberalism.* Oxford: Oxford University Press.

Hayek, F. A. v. 1944. *The Road to Serfdom.* London: Routledge.

Hayles, N. K. 2005. *My Mother Was a Computer: Digital Subjects and Literary Texts.* Chicago: University of Chicago Press.

Hirschman, A., O. 1970. *Exit, Voice, and Loyalty: Responses to Decline in Firms, Organizations, and States.* Cambridge, MA: Harvard University Press.

Hobbes, T. 1985. *Leviathan.* London: Penguin Books.

Holloway, T., Bozicevic, M., and Börner, K. 2005. "Analyzing and Visualizing the Semantic Coverage of Wikipedia and Its Authors." *Complexity* 12 (3): 30–40.

Howe, J. 2006. "The Rise of Crowdsourcing." *Wired* 14 (6). Retrieved from http://www.wired.com/wired/archive/14.06/crowds.html.

Huhtamo, E., and J. Parikka, eds. 2011. *Media Archaeology: Approaches, Applications, and Implications.* Berkeley: University of California Press.

Innis, H. A. 1950. *Empire and Communications.* Oxford: Oxford University Press.

———. 1951. *The Bias of Communication.* Toronto: University of Toronto Press.

Jenkins, H. 1992. *Textual Poachers: Television Fans and Participatory Culture.* New York: Routledge.

———. 2006a. *Convergence Culture: Where Old and New Media Collide.* New York: New York University Press.

———. 2006b. *Fans, Bloggers, and Gamers: Exploring Participatory Culture.* New York: New York University Press.

———. 2009. *Confronting the Challenges of Participatory Culture: Media Education for the 21st Century.* Cambridge, MA: MIT Press.

Keating, P., and A. Cambrosio. 2003. *Biomedical Platforms: Realigning the Normal and the Pathological in Late Twentieth-century Medicine.* Cambridge, MA: MIT Press.

Kelty, C. M. 2008. *Two Bits: The Cultural Significance of Free Software.* Durham, NC: Duke University Press.

Kendall, W. 1960. "The 'Open Society' and Its Fallacies." *American Political Science Review* 54 (4): 972–79.

Kildall, S., and N. Stern. 2011. "Wikipedia Art: Citation as Performative Act." In *Critical Point of View: A Wikipedia Reader*, ed. G. Lovink and N. Tkacz, 165–90. Amsterdam: Institute of Network Cultures.

King, J. 2006. "Openness and Its Discontents." In *Reformatting Politics: Information Technology and Global Civil Society*, ed. J. Dean, J. W. Anderson and G. Lovink, 43–54. New York: Routledge.

Kirschenbaum, M. G. 2008. *Mechanisms: New Media and the Forensic Imagination.* Cambridge, MA: MIT Press.

Kittler, F. A. 1990. *Discourse Networks 1800/1900.* Stanford, CA: Stanford University Press.

———. 1997. *Literature, Media, Information Systems: Essays.* Edited by J. Johnston. Amsterdam: G and B Arts International.

———. 1999. *Gramophone, Film, Typewriter.* Translated by G. Winthrop-Young and M. Wutz. Stanford, CA: Stanford University Press.

———. 2010. *Optical Media: Berlin Lectures 1999.* Translated by A. Enns. Cambridge: Polity Press.

Kittur, A., E. Chi, B. Pendleton, B. Sun, and T. Mytkowicz. 2007. "Power of the Few vs Wisdom of the Crowd: Wikipedia and the Rise of the Bourgeoisie." Paper presented at the 25th Annual ACM Conference on Human Factors in Computing Systems (CHI 2007), San Jose, CA.

Kittur, A., and R. Kraut. 2008. "Harnessing the Wisdom of Crowds in Wikipedia." Paper presented at the ACM 2008 Conference on Computer Supported Cooperative Work, New York.

Knabb, K., ed. 1981. *Situationist International Anthology.* Berkeley, CA: Bureau of Public Secrets.

Konieczny, P. 2010. "Adhocratic Governance in the Internet Age: A Case of Wikipedia." *Journal of Information Technology and Politics* 7:263–83.

Kostakis, V. 2010. "Identifying and Understanding the Problems of Wikipedia's Peer Governance: The Case of Inclusionists versus Deletionists." *First Monday* 15 (3).

Lakoff, G. 2002. *Moral Politics: How Liberals and Conservatives Think.* Chicago: University of Chicago Press.

———. 2004. *Don't Think of an Elephant! Know Your Values and Frame the Debate.* White River Junction, VT: Chelsea Green Publishing.

Lanier, J. 2006. "Digital Maoism: The Hazards of the new Online Collectivism." *Edge—The Third Culture.* Retrieved from http://www.edge.org/3rd_culture/lanier06/lanier06_index.html.

Lathrop, D., and L. Ruma, eds. 2010. *Open Government: Collaboration, Transparency, and Participation in Practice.* Sebastopol, CA: O'Reilly Media.

Latour, B. 1987. *Science in Action: How to Follow Scientists and Engineers through Society.* Cambridge, MA: Harvard University Press.

———. 1991. "Technology Is Society Made Durable." In *A Sociology of Monsters*, ed. J. Law, 103–31. London: Routledge.

———. 1992. "Where Are the Missing Masses? The Sociology of a Few Mundane Artifacts." In *Shaping Technology/Building Society: Studies in Sociotechnical Change*, ed. W. Bijker and J. Law, 225–58. Cambridge, MA: MIT Press.

———. 1996. *Aramis, or, The Love of Technology*. Cambridge, MA: Harvard University Press.

———. 1999. *Pandora's Hope: Essays on the Reality of Science Studies*. Cambridge, MA: Harvard University Press.

———. 2002. "There Is No Information, Only Transformation." In *Uncanny Networks: Dialogues with the Virtual Intelligentsia*, ed. G. Lovink, 154–61. Cambridge, MA: MIT Press.

———. 2005. *Reassembling the Social: An Introduction to Actor-Network-Theory*. Oxford: Oxford University Press.

———. 2008. "What's Organizing? A Meditation on the Bust of Emilio Bootme in Praise of Jim Taylor." Paper presented at the Materiality, Agency and Discourse Conference 2008, HEC Montréal. Retrieved from http://youtu.be/rrMxyQP3rYY.

Latour, B., and S. Woolgar. 1986. *Laboratory Life: The Construction of Scientific Facts*. Princeton, NJ: Princeton University Press.

Law, J. 2004. *After Method: Mess in Social Science Research*. London: Routledge.

Leonard, A. 2013a. "Wikipedia's Shame." *Salon*. Retrieved from http://www.salon.com/2013/04/29/wikipedias_shame/.

———. 2013b. "Revenge, Ego and the Corruption of Wikipedia." *Salon*. Retrieved from http://www.salon.com/2013/05/17/revenge_ego_and_the_corruption_of_wikipedia/.

———. 2013c. "Wikipedia Cleans Up Its Mess." *Salon*. Retrieved from http://www.salon.com/2013/05/21/wikipedia_cleans_up_its_mess/.

Lessig, L. 2005. "Open Code and Open Societies." In *Perspectives on Free and Open Source Software*, ed. J. Feller, B. Fitzgerald, S. A. Hissam and K. R. Lakhani, 349–60. Cambridge, MA: MIT Press.

———. 2008. *Remix: Making Art and Commerce Thrive in the Hybrid Economy*. New York: Penguin Press.

Levinson, R. B. 1970. *In Defense of Plato*. New York: Russell and Russell.

Levy, S. 1984. *Hackers: Heroes of the Computer Revolution*. Garden City, NY: Anchor Press/Doubleday.

Lih, A. 2009. *The Wikipedia Revolution: How a Bunch of Nobodies Created the World's Greatest Encyclopedia*. New York: Hyperion.

Locke, J. 1976. *The Second Treatise of Government*. 3rd ed. Oxford: B. Blackwell.

Lovink, G., and N. Tkacz. 2011. *Critical Point of View: A Wikipedia Reader*. Amsterdam: Institute of Network Cultures.

Lyotard, J.-F. 1988. *The Differend: Phrases in Dispute*. Translated by G. V. D. Abbeele. Minneapolis: University of Minnesota Press.

———. 1993. *Political Writings*. Translated by B. Readings. London: UCL Press.

Magee, B. 1982. *Popper*. London: Fontana.

Manovich, L. 2013. *Software Takes Command*. New York: Bloomsbury Publishing.

Marcuse, H. 1991. *One-Dimensional Man: Studies in the Ideology of Advanced Industrial Society*. Boston: Beacon Press.

Mirowski, P. 2009. "Postface: Defining Neoliberalism." In *The Road from Mont Pèlerin: The Making of the Neoliberal Thought Collective*, ed. P. Mirowski and D. Plehwe, 417–55. Cambridge: Harvard University Press.

———. 2013. *Never Let a Serious Crisis Go to Waste: How Neoliberalism Survived the Financial Meltdown*. London: Verso.
Mirowski, P., and D. Plehwe. 2009. *The Road from Mont Pèlerin: The Making of the Neoliberal Thought Collective*. Cambridge, MA: Harvard University Press.
Mol, A. 2002. *The Body Multiple: Ontology in Medical Practice*. Durham, NC: Duke University Press.
Morozov, E. 2013. *To Save Everything, Click Here: Technology, Solutionism, and the Urge to Fix Problems that Don't Exist*. London: Allen Lane.
Mouffe, C. 1993. *The Return of the Political*. New York: Verso.
———. 2005. *On the Political*. New York: Routledge.
Niederer, S., and J. van Dijck. 2010. "Wisdom of the Crowd or Technicity of Content? Wikipedia as a Sociotechnical System." *New Media and Society* 12 (8): 1368–87.
Nozick, R. 1974. *Anarchy, State, and Utopia*. New York: Basic Books.
O'Neil, M. 2009. *Cyberchiefs: Autonomy and Authority in Online Tribes*. London: Pluto Press.
———. 2011. "Wikipedia and Authority." In *Critical Point of View: A Wikipedia Reader*, ed. G. Lovink and N. Tkacz, 309–24. Amsterdam: Institute of Network Cultures.
O'Reilly, T. 2010. "Government as a Platform." In *Open Government: Collaboration, Transparency, and Participation in Practice*, ed. D. Lathrop and L. Ruma, 11–39. Sebastopol, CA: O'Reilly Media.
Ong, W. J. 2002. *Orality and Literacy: The Technologizing of the Word*. London: Routledge.
Papadopoulos, D., N. Stephenson, and V. Tsianos. 2008. *Escape Routes: Control and Subversion in the 21st Century*. London: Pluto Press.
Parikka, J. 2007. *Digital Contagions: A Media Archaeology of Computer Viruses*. New York: Peter Lang.
———. 2010. *Insect Media: An Archaeology of Animals and Technology*. Minneapolis: University of Minnesota Press.
———. 2012. *What Is Media Archaeology?* Cambridge: Polity Press.
Pasquinelli, M. 2008. *Animal Spirits: A Bestiary of the Commons*. Rotterdam: NAi Publishers; Amsterdam: Institute of Network Cultures.
Plato. 1974. *The Republic*. Translated by H. D. P. Lee. 2nd ed. Baltimore, MD: Penguin.
Poe, M. 2006. "The Hive." *Atlantic Online* (September). http://www.theatlantic.com/doc/200609/wikipedia.
Popper, K. R. 1962. *The Open Society and Its Enemies*. 2 vols. 4th ed. London: Routledge and Kegan Paul.
Reagle, J. M. 2005. "A Case of Mutual Aid: Wikipedia, Politeness, and Perspective Taking." *Proceedings of Wikimania 2005—The First International Wikimedia Conference, Frankfurt, Germany*. http://reagle.org/joseph/2004/agree/wikip-agree.html.
———. 2007. "Do As I Do: Authorial Leadership in Wikipedia." Paper presented at the WikiSym '07, Montre al, Quebec, Canada. http://ws2007.wikisym.org/space/ReaglePaper/Reagle_WikiSym2007_WikipediaAuthorialLeadership.pdf.
———. 2008. "In Good Faith: Wikipedia Collaboration and the Pursuit of the Universal Encyclopedia." PhD dissertation, New York University, New York.
———. 2010. *Good Faith Collaboration: The Culture of Wikipedia*. Cambridge, MA: MIT Press.
Rheingold, H. 2002. *Smart Mobs: The Next Social Revolution*. Cambridge, MA: Perseus Publishing.

Rousseau, J.-J. 1986. *The Social Contract and Discourses.* Translated by J. C. Hall, G. D. H. Cole, and J. H. Brumfitt. London: Dent.

Rushkoff, D. 2003. *Open Source Democracy: How Online Communication Is Changing Offline Politics.* London: Demos. http://rushkoff.com/books/open-source-democracy/.

Sanger, L. 2000. "Epistemic Circularity: An Essay on the Problem of Meta-justification," PhD diss., Ohio State University. http://enlightenment.supersaturated.com/essays/text/larrysanger/diss/preamble.html.

———. 2002e. "Re: Good luck with Your WikiPaidia." http://osdir.com/ml/science.linguistics.wikipedia.international/2002-02/msg00039.html.

Schutz, A. 1970. *On Phenomenology and Social Relations: Selected Writings.* Chicago: University of Chicago Press.

Serres, M. 2007. *The Parasite.* Translated by L. R. Schehr. Minneapolis: University of Minnesota Press.

Shearmur, J. 1996. *The Political Thought of Karl Popper.* New York: Routledge.

Shirky, C. 2008. *Here Comes Everybody: The Power of Organizing without Organizations.* New York: Penguin Press.

Siefkes, C. 2008. *From Exchange to Contributions: Generalizing Peer Production into the Physical World.* Berlin: Siefkes-Verlag.

Spek, S., E. Postma, and H. J. van den Herik. 2006. "Wikipedia: Organisation from a Bottom-up Approach." Paper presented at the WikiSym 2006, Odense, Denmark. http://arxiv.org/abs/cs.DL/0611068.

St. Laurent, A. M. 2004. *Understanding Open Source and Free Software Licensing.* Cambridge, MA: O'Reilly Media.

Stalder, F., and J. Hirsch. 2002. "Open Source Intelligence." *First Monday* 7 (6). http://firstmonday.org/htbin/cgiwrap/bin/ojs/index.php/fm/article/view/961/882.

Stvilia, B., M. B. Twidale, L. Gasser, and L. C. Smith. 2005. "Information Quality in a Community-Based Encyclopedia." Paper presented at the International Conference on Knowledge Management, Hackensack, NJ.

Sunstein, C. R. 2006. *Infotopia: How Many Minds Produce Knowledge.* Oxford: Oxford University Press.

Surowiecki, J. 2004. *The Wisdom of Crowds: Why the Many Are Smarter than the Few and How Collective Wisdom Shapes Business, Societies and Nations.* New York: Doubleday.

Tapscott, D., and A. D. Williams. 2006. *Wikinomics: How Mass Collaboration Changes Everything.* New York: Penguin.

Tkacz, N. 2010. "Wikipedia and the Politics of Mass Collaboration." *Platform: Journal of Media and Communication* 2 (2): 40–53.

———. 2011. "The Spanish Fork: Wikipedia's Ad-fuelled Mutiny." *Wired.co.uk.* http://www.wired.co.uk/news/archive/2011-01/20/wikipedia-spanish-fork?page=all.

Toffler, A. 1970. *Future Shock.* New York: Random House.

Trotsky, L. 1954. "It Is Necessary to Drive the Bureaucracy and the New Aristocracy Out of the Soviets." *Fourth International* 15 (1): 34–35.

Vernon, R. 1976. "The 'Great Society' and the 'Open Society': Liberalism in Hayek and Popper." *Canadian Journal of Political Science* 9 (2): 261–76.

Virno, P. 1996. "Virtuosity and Revolution: The Political Theory of Exodus." In *Radical Thought in Italy: A Potential Politics,* ed. P. Virno and M. Hardt, 189–212. Minneapolis: University of Minnesota Press.

Virno, P., and Costa, F. (n.d.). "Between Disobedience and Exodus: Interview with Paolo Virno with Flavia Costa." *Generation Online*. Retrieved January 10, 2012, from http://www.generation-online.org/p/fpvirn05.htm.

Vismann, C. 2008. *Files: Law and Media Technology.* Stanford, CA: Stanford University Press.

von Bertalanffy, L. 1950. "The Theory of Open Systems in Physics and Biology." *Science* 111:23–29.

———. 1960. *Problems of Life: An Evaluation of Modern Biological and Scientific Thought.* New York: Harper and Brothers.

Walzer, M. 1985. *Exodus and Revolution.* New York: Basic Books.

Weber, M. 1958. *From Max Weber: Essays in Sociology.* Translated by H. H. Gerth and C. W. Mills. New York: Oxford University Press.

Weber, S. 2004. *The Success of Open Source.* Cambridge, MA: Harvard University Press.

Whyte, W. H. 1956. *The Organization Man.* New York: Simon and Schuster.

Williams, R. 1976. *Keywords: A Vocabulary of Culture and Society.* London: Fontana.

Williams, S. 2002. *Free as in Freedom: Richard Stallman's Crusade for Free Software.* Sebastopol, CA: O'Reilly Media.

Yeo, R. 2001. *Encyclopaedic Visions: Scientific Dictionaries and Enlightenment Culture.* Cambridge: Cambridge University Press.

Zielinski, S. 2006. *Deep Time of the Media: Toward an Archaeology of Hearing and Seeing by Technical Means.* Cambridge, MA: MIT Press.

Index

actor-network theory (ANT), 10, 73, 114, 156, 157n9, 169
ad-hocracy, 7, 9, 37–41, 88, 91–97, 124–26, 159, 179–81
advertising (ads), 150–56, 160–66, 169–75
agency, 9, 17, 20–21, 29, 33, 36, 38, 43, 46, 72, 78, 117, 119, 122–24, 134, 141, 158–59, 170–72
agonism, 7, 10, 136
Aguerriero, 68–69, 196–97
Alexa, 148
allies, 32–33, 37, 107, 110, 146, 154, 156–57, 161, 163–76
anonymous (users), 9, 65–66, 113
antagonism, 7, 33, 36, 136
anyone can edit, 6, 11, 28, 44, 50, 98, 100–101
Apache, 25, 145
apparatus of material implements, 9, 98, 111–12, 115, 117, 119, 124, 181
Arbitration Committee, 69
archaeology, 4, 38–40, 98n8
archives, 24, 45, 55, 66–70, 79n33, 82, 98–99, 105n20, 183–97
aristocracy, 88, 93, 206
Aristotle, 17–18
artful juggling, 81
articles for deletion (AfD), 53–59, 76
Assange, Julian, 3
Associative Man, 91–92, 96–97, 126
asymmetry, 7, 36, 172, 175
Austin, John L., 39, 118
authority, 6, 7, 85–86, 93, 97–98, 104, 110, 119–24, 128, 134, 159–60, 172, 176, 178–80
axioms, 70–71

Bateson, Gregory, 8, 70–72
Bauwens, Michel, 93, 137–38

bazaar. *See* "Cathedral and the Bazaar, The"
benevolent dictator, 6, 93–94, 165
Benkler, Yochai, 4, 44–49, 83, 85
Bennis, Warren, 90n2
Bergson, Henri, 15
bias, 62, 66–67, 78, 104–8
black box, 166
Bomis, 146, 150–53, 162–66, 169–72
Bot Approval Group (BAG), 113–14
bot policy, 113–14
bots, 9, 10, 111–18, 125, 145, 148, 180
boundary objects, 9, 81–86, 181
boundary statements, 77, 81–84
Bowker, Geoffrey, 9, 81
boyd, danah, 62
Bruns, Axel, 6, 43–49, 84, 86, 93, 112n36
bureau, the, 97–98, 111, 124–25
bureaucracy, 9, 88–98, 110–11, 124, 126, 179–81

cabal, the, 93
Cage, John, 51
Callon, Michel, 73, 156–57, 160
capitalism, 5, 18, 33, 36, 89, 95, 130–31, 178
Carr, Nicholas, 92–94
Carswell, Douglas, 31
"Cathedral and the Bazaar, The," 2, 23–24, 30–31, 46, 180
Catholic Church, 94, 97
cave, simile of the, 16
censorship, 63–68, 75–76, 80, 82, 123, 183–84, 189–94
centralization, 20, 23, 30, 44, 48, 176–81
centralized planning, 18–19, 24
Chambers, Ephraim, 100
change, 16–18, 36, 89–98
charisma, 7

Chun, Wendy Hui Kyong, 141–42, 144, 146, 180
Citizendium, 171n37, 174n41
class struggle, 17, 34
closed, 15, 18, 24, 33, 36, 178, 180–81; code, 14; decision making, 30; infrastructures, 20; society, 16–17, 34, 178–79; source, 37; system, 15, 37; thought, 17
code, 14, 21–23, 30, 36, 111, 114, 118, 131–39, 141–43, 155, 158
Coleman, Gabriella, 23
collaboration, 1–2, 6–8, 11, 25–27, 29–32, 37–38, 40–50, 63, 70, 83–86, 112n36, 126, 138, 153, 181; competition, 179; mass, 6, 112n6, 138; neutrality, 107–8; sorting, 8, 42, 49–50
commons-based peer production. *See* peer production, commons-based
communism, 17–18, 33, 36, 88–89
Communist Party, 89
community, 22, 24, 30, 47–48, 53–54, 81, 91, 93, 120, 131–33, 135, 137, 144–48, 150–51, 153, 155, 158–59, 163, 166, 170, 172–74
competition, 18, 20, 29–30, 32, 34–36, 67, 131, 147, 174, 179
computation, regime of, 126, 139–40, 146, 148
computationalism, 139–40, 146
concern, matter of, 152
conscious control, 19
consensus, 50, 53–54, 58, 68n26, 68–70, 76, 80, 94, 101, 106, 108, 113, 136–37, 148, 158, 187, 189, 191; policy (Wikipedia), 75; politics, 136n13, 158n10
constitutive outside, 136n13
Contract from America (Tea Party), 31
controversy, 7, 9–10, 12, 50, 63–64, 67, 76, 103–4, 116, 153, 160, 164–69, 171–74, 176, 179
copyright, 4, 22, 59, 131
Cramer, Florian, 105–6
criteria for speedy deletion, 53, 59, 61, 74
crowd, 157–59, 163, 167, 169, 172, 173
crowds, wisdom of the, 112, 116
cybernetics, 3, 32, 126

d'Alembert, Jean le Rond, 100
damages, 77–78, 86, 136, 181
DanielRigal, 56–58, 60–61, 75
Danish cartoon controversy, 63, 76, 80, 188
Darnton, Robert, 4–5, 10
decentralization, 19, 44, 83, 85, 176, 179
de Certeau, Michel, 42
delegation, 114
deletion, request for, 65–66, 70, 76
Deleuze, Gilles, 39–40
democracy, 1–2, 14, 16, 18, 29–31, 35–36, 88–89, 93–94, 136, 138–39
Derrida, Jacques, 39, 136
diagram, 70–71, 81, 140
dictatorship, 35, 93–94, 136, 165

Diderot, Denis, 4, 100
differend, 9, 77–86, 181
Digital Maoism, 138
discourse, 3, 9, 40, 74n30, 77, 84, 87, 98, 100, 110, 115–16, 127, 139–40, 146, 149, 151, 157–58, 166–67, 176
discourse analysis, 38
discussion lists, 150–51
discussion pages, 54, 64, 76, 112–13
dissent assemblage, 169, 176
dissenter, 152, 157, 161, 166–69, 172–76, 181
division of labor, 45, 48, 83, 85, 90
donations, 3, 161
durability, 41, 50, 73, 115, 124, 147, 153, 168, 174, 176, 181
dwarfism, 167, 169

Eco, Umberto, 51
edit, anyone can. *See* anyone can edit
edit requests, 65–66, 70
edit wars, 62–64, 101
elections, 31, 35, 120, 129, 156
Emergent Bureaucracy, 92. *See also* Carr, Nicholas
Enciclopedia Libre Universal en Español (EL), 144–48, 151
encyclopedia 3, 6–8, 10–11, 28, 49, 54, 56, 59–60, 62–64, 67, 75, 77, 80, 84–86, 100–101, 105, 108–10, 119, 152; history, 4–5
Encyclopédie, 4, 5, 10
Enlightenment, 4, 5, 10, 76
entomology, 45
Enyedy, Edgar, 144–47, 150–63, 165–66, 168–72, 174–75, 176
escape, 126
exchange, 4, 30, 34–36, 131
Exclusion Compliant, 116, 123
ex corpore, 88, 124–25, 181
exit, 8–10, 41, 126, 129–30, 134–35, 138–39, 146, 148, 173, 176, 179, 181. *See also* forking
exodus, 126
Experimental GNU Compiler System (EGCS), 141

fans, 43
fascism, 17–19, 33, 36, 88–89
feedback, 130, 141
fetish, 142, 146
files, 97–98, 110–11, 120, 122, 181. *See also* script
Filipacchi, Amanda, 11–12
firms, 44, 83, 85, 90, 129–30
fishermen, 156–57
Five Pillars, 56, 60, 74, 98, 100–103, 106
forking, 7–10, 37, 41, 126–27, 130–39, 148–49, 163, 179, 181; computationalism, 139–44, constitutive, 146–48; 133–34, 136, 139; controversy, 10, 150–76; lossless, 40–41, 138–39, 144, 148, 173; safety net, 133–39, 181; statement formations,

173–76. *See also* Enciclopedia Libre Universal en Español; exit; Spanish Wikipedia
forensic materiality, 143–44, 146, 148–49, 154n7
formal equivalence, working model of, 146–49, 154n7, 163, 174–75, 181
formal materiality, 143–44, 146–49, 154n7, 174
Foucault, Michel, 38–40, 43, 154, 177n1, 178
FOSS. *See* free and open source software
FLOSS. *See* free and open source software
frame, 8–9, 70–87, 100, 117, 121–25, 136, 152, 176, 182; alignment, 74; dispute, 74; exception, 80; wars, 75
Frame Analysis, 72–74
Frankfurt School, 89
free and open source software (FLOSS/FOSS), 7, 25, 101, 125, 131n8, 141
free beer, 155
freedom, 19–20, 22, 24, 29, 35–36, 79, 123n46, 126–27, 133–35, 137–39, 155–56, 166, 170
free software, 14, 20, 23, 126, 131n8, 133, 135–37, 155–56
Free Software Foundation (FSF), 22–23, 26–27
Free Software Magazine, 135
Free Software Movement (FSM), 22–24, 132n10, 141, 165
French Encyclopedia. *See* Encyclopédie
Frequently Asked Questions (FAQ), 64–66, 68, 70, 76, 161
future shock, 89–90, 97

Galloway, Alexander, 118
Gardner, Sue, 12
Geiger, R. Stuart, 111–14, 116, 119
genealogy, 40, 125
genre, 4, 62–63, 67, 77–78, 100, 119, 155, 160
German Nationalism, 17
global financial crisis, 1, 36, 177
GNU C Compiler (GCC), 141
GNU Free Documentation License, 4n4, 28, 145
GNU General Public License (GNU GPL), 22–23, 23n11
GNU operating system, 22, 25
Goffman, Erving, 8, 72–74, 76
Golumbia, David, 140, 142
Google, 4, 23n12, 25, 28, 47
Google Search, 11, 47
governance, 2, 7, 9, 20, 26, 28, 32–33, 35–36, 40, 46, 86, 88–125, 127, 133–35, 137, 139, 148, 174, 176, 181
governmentality, 180
Government 2.0 Taskforce (Australia), 31
granularity, 4, 45–46
Greek tribalism, 15–16
Guattari, Felix, 39–40

HagermanBot, 111–19
Hancock, Terry, 135–37

happiness, 88
Hardt, Michael, 14, 37
hardware, 29, 131, 180
Hayek, Friedrich, 15n1, 18n4, 19–20, 24, 30, 37, 44, 105, 178
Hayles, N. Katherine, 118, 139–40, 146
Hegel, Georg Wilhelm Friedrich, 17–18
hierarchy, 6, 9, 44–50, 67, 84–86, 90–92, 96–97, 120, 126, 136n13, 179–81
hinterland, 76, 94, 100–101, 104–5, 111, 125, 168, 173
Hirschman, Albert, 9, 126–27, 129–30, 134, 138
historicism, 17
Hobbes, Thomas, 127–29

ideal form, 16, 88
images. *See also* Muhammad
individualism, 34–35, 140
Innis, Harold, 42
interface, 142
intermediaries, 157–58
inscription, 39, 100–101, 119, 125, 142–43, 156n8, 165
inscription machines, 39
institutional politics, 3, 14, 29, 32
instructuring, 101
instructuring agency, 122–23
IP address, 113
Islam, 63, 65, 67–68, 75, 80, 183–85, 188–90, 194

janitor, 121
javelins, rusty, 96–97
Jenkins, Henry, 6, 42–43
Jesus, 67, 107, 109, 190, 193
justice, 34–35, 88–89
Jyllands-Posten, 63

karma, 47
Kelty, Christopher, 20, 22, 24, 37, 131–32
keys, hotel, 115, 118, 157
Kildall, Scott, 8, 50–56. *See also* Wikipedia Art
Kirschenbaum, Matthew, 143–44, 146

lamest edit wars, 62
Lanier, Jaron, 138
Latour, Bruno, 10, 39–40, 40n34, 73, 99, 114–15, 117, 122–24, 143, 152–60, 166–68, 173, 176
Law, John, 40n34, 76
leadership, 86, 93, 137, 158–60, 176, 179, 181
left, the, 177
legitimating mechanisms, 9–10, 88–89, 94–95, 127, 130, 134, 139, 147, 175, 181, 182
Lessig, Lawrence, 14, 22, 29, 37
Leviathan, 128
liberal democracy, 33, 35, 88
liberalism, 19, 24, 32, 127, 140
liberogenic, 178
libertarian, 31–32, 127n1

Linux operating system, 23–25
Locke, John, 129n3
loyalty, 129. *See also* Hirschman, Albert
LWN.net, 132
Lyotard, Jean-François, 9, 77–78, 82, 85, 136

"magic circle," 11
"making do," 81
managerial commands, 44, 46, 85, 126
Marcuse, Herbert, 89
markets, 2, 8, 19–20, 24, 32, 35–36, 44, 46, 48, 83–85, 91, 126, 129, 134–35, 139, 178–81
Marx, Karl, 17–18, 32, 149
materiality, 5, 10, 73, 111, 115, 154, 157. *See also* forensic materiality; formal materiality
MeatballWiki, 131, 133
mediation, 43, 118, 136
mediation (Wikipedia procedure), 8, 66, 68–70, 76, 85, 193–97
mediator, 158
merit, 7, 49–50. *See also* meritocracy
meritocracy, 45, 48, 84, 86, 93, 94, 137, 139
meta-communication, 70, 72, 80, 84
method, 8, 12, 38, 40, 73, 125, 143, 181
Microsoft, 22n10, 163, 165
Miranda, Joao, 165
Mirowski, Philip, 20n6, 177–78
MIT (Massachusetts Institute of Technology), 20, 22, 26, 36
modal relations, 155–56, 159–60, 162, 168–69, 176
modularity, 4, 45–46, 63, 96
MOOCs (Massive Open Online Courses), 26
Mouffe, Chantal, 7, 136, 168
mouthpiece, 157–60, 163, 166, 169–72, 176, 179, 181. *See also* spokesperson
Mozilla, 25
Mozilla Public License, 23
Mr Anybody, 167
Mr Manybodies, 167
Mr Somebody, 167–68
Muhammad, 8, 50, 61, 63–70, 74–85, 99, 103, 121, 123, 183–97
multitude, 14, 97, 128

natural rights, 129
Nazi Party, 97
Negri, Antonio, 14, 37, 130n4
neoliberalism, 18–20, 32, 35–37, 62, 177–81
network cultures, 2, 3, 10, 14, 18, 28–29, 32, 39, 126, 151
neutrality, 64, 66, 104–5, 107–9, 125
neutral point of view (NPOV), 9, 49, 62, 64–65, 67, 74, 76, 86, 100, 104–10, 121, 125, 173, 181, 186–87, 195, 197
Newsnight, 31

New York Times, 12, 31
nondisclosure agreement (NDA), 21, 36, 115
nonhuman, 10, 70, 71, 114
no original research (NOR), 61, 64, 74, 106–7, 110
notability, 57
Nozick, Robert, 127n1
Nupedia, 93n5, 104–5, 107, 163–64, 171

Obama, 1, 3, 7, 31
Objectivist philosophy, 105
office, the, 97, 111
officials, 97–98, 111, 119, 124
open: access, 14, 25–26; community, 14, 133; education, 14, 25–26; government, 1–3, 29–35, 37, 180; participation, 6, 45, 105; platform, 2, 31, 180; politics, 14–38; project, 9–10, 26, 38, 40, 47, 84, 132, 174, 176, 181; society, 8, 14–19, 30, 33–37; source software, 2, 7, 14, 18, 20–21, 23, 25, 27–31, 101, 112n36, 131–34, 137, 141; systems, 7, 14–15, 18, 20, 24, 37
Open Courseware Consortium, 26
Open Everything, 26–27, 37
Open Government Initiative, 2, 3, 31
Open Humanities Press, 25
Open Knowledge Foundation, 26
OpenLeaks, 3
OpenOffice.org, 135
Open Source Initiative (OSI), 23–24, 29, 132n10, 141
operating system (OS), 20, 22–25, 111, 130, 141
order words, 39–40
O'Reilly, Tim, 2, 30
organization, 7, 10, 12–13, 16, 18–20, 20n6, 24–25, 30, 36–38, 40, 44–46, 48, 72–73, 83–99, 111, 117, 119, 122, 124–30, 134, 136, 138–39, 146–48, 159, 164, 177–82
organizational charts, 90
organizational man, 90–91, 96

Parasite, The, 154n7
parliament of things, 10
participation, 2, 6, 7, 8, 11, 26–27, 29–32, 37–38, 40, 42–44, 49, 51, 70, 83, 105, 126, 158
participatory culture, 6, 43
peer production, commons-based, 43–44, 46, 133, 141
Peloponnesian War, 15, 18
perception, 16, 29, 71
performativity, 39, 52, 71, 117–19, 124, 139, 173
philosophes, 100
picture frame, 71
PIPA (Protect IP Act), 10
platform, 2, 30–31, 56–57, 125, 140, 180
Plato, 16–18, 33–34, 94
play, 70–74, 92, 96
Pliny, 100

"Policies and guidelines, List of," 102–3
Polish Wikipedia, 162
political, the, 7, 136, 168
politics, institutional, 3, 14, 29, 32
poll, 69. *See also* vote
Popper, Karl, 8, 14–20, 30, 32–37, 44, 88, 178
posthuman, 156
postpolitical, 7
precarious workers, 97
price mechanism, 44, 46, 126, 155
principles, 6, 9, 25, 31, 39, 49, 56, 70–71, 76, 78, 80, 84–87, 94, 98–102, 106, 108–9, 117, 129, 133, 156, 182
produsage, 45, 47–49, 63, 112n36
project manager, 90
proletariat, 17, 89
property, 24, 36–37, 48, 101, 155
proprietary, 20, 22, 24, 44, 137, 174
protected pages, 45, 66, 120
P2P Alternatives, Foundation for, 27, 32
P2P civilization, 137

quasi-object, 154n7
Quran, 77, 80, 183–84
Qworty, 12

Rand, Ayn, 105
Raymond, Eric, 2, 23–24, 29–30, 44, 48, 131
Reagle, Joseph, 6, 43, 49, 86, 93, 105, 107–8, 133
reason, 34–35, 55, 70, 125
relic, 98, 125
rent, 180
Republic, The, 16, 88, 93–94
revert wars, 64
revolution, 9, 35, 52n4, 89, 92, 127, 129, 135, 151n4
Rheingold, Howard, 43
roles, 16–17, 48, 73, 90, 97–98, 120–21
Rousseau, Jean-Jacques, 127–28n2
Rozeta wiki, 162n16, 164
rules, 6, 40, 44, 47–49, 53, 59, 61, 73, 77, 92, 97–102, 107, 110, 121–23, 128, 148, 152, 171, 173, 184

safety net. *See* forking
Sanger, Larry, 93n5, 104–6, 150–55, 159–76
scale, 147–49, 181
scallops, 156–57
scarcity, 22, 131
Schutz, Alfred, 70
science, 1, 5, 10, 18n4, 26, 29, 44, 91, 143
science and technology studies (STS), 10, 39, 153–55, 157n9, 160, 166–68
Science in Action, 154, 166
scribes, 97, 119, 181
script, 122–23
script (code), 111, 145

self-organization, 6, 46, 84, 94
Serres, Michel, 154n7
set theory, 70
Shirky, Clay, 6, 31, 43–45, 49, 84–86, 112
signals, 44, 70, 85, 106, 181
SineBot, 112
Situationists, 89
Slashdot, 47, 93n5, 104
"sleeping police officers," 114
social, the, 136
social contract, 127–28n2
social engineer, 17, 33–34
sociology, 12, 70, 88, 157n9
software cultures, 2, 4, 18, 20, 23, 25, 30, 32, 93, 126, 130
SOPA (Stop Online Piracy Act), 10
source code, 14, 21, 32, 36, 131, 134, 141–42, 155
Sourceforge, 145
sourcery, 142, 144, 146, 148, 175, 181
sovereignty, 127–29
Spanish Fork, 9–10, 127, 144, 147, 174–75. *See also* Enciclopedia Libre Universal en Español; forking
Spanish Wikipedia, 9, 144–66, 169–76
speech acts, 38–39, 118n39
speed bump, 114
spokesperson, 156–60, 163, 166, 169, 171–73, 176, 181
sponsorships, 161–62, 164
Stallman, Richard, 20–25, 30, 36, 141, 156
Star, Susan Leigh, 9, 81
statement formations, 40–41, 87, 124–25, 153, 173–74
Statement of Principles, 101
statements, 9–10, 38–41, 67–68, 70, 75–88, 98–141, 147, 151–62, 167–70, 172–76, 181; boundary object-tending 83; differend-tending, 79–80; forceful, 41, 85, 88, 99, 101, 106, 107n26, 110, 117, 124–25, 159, 172–73, 176, 181; framing, 9, 77, 121–25; games, 85, 152, 171, 181; loading, 99, 115–16, 118, 154, 157, 173–75, 176, 181; modalities, 152–56, 160, 164, 166, 168–69, 172, 176, 181 (*see also* modal relations); plane of, 82–83; political ontology of, 10
state of nature, 127
State of Wikipedia, The, 102
St. Brieuc Bay, 156
Stern, Nathaniel, 8, 50–51, 54, 56
stigmergy, 46, 181. *See also* collaboration
strength, trial of, 117, 173
structure of experience, 72–73
sysop, 10, 120–21

task forces, 31, 45, 90–91, 95, 158n10
TCP/IP, 20
Tea Party, 31, 37

templates, 113
"textual poachers," 42
theatrical frame, 73
tinge, neoliberal, 177, 179
Toffler, Alvin, 89–97, 124
Torvalds, Linus, 23–24
totalitarianism, 16, 19–20, 35–37, 89, 177–78, 180
transformative agency, 123
translation, sociology of, 157n9
transparency, 1–2, 25–26, 29, 31–32, 37
Trifecta, 101
Tron, 118
Trotsky, Leon, 89
truth, 3, 17–18, 34, 36, 39, 40, 49, 51–52, 55, 57, 64, 79, 100–101, 105–10, 125, 139, 142, 178–79
Turing Machine, Universal, 140
tyranny, 35, 129n3, 136, 139, 181

UNIX operating system, 20, 22–23
unsigned, 112–14, 116, 118
Unsigned template, 113–14, 116
UseModWiki, 145
user-access levels. *See* user hierarchies
user hierarchies, 9, 120–24, 171n37
username, 113

vandalism, 53, 59, 64, 74–75, 112
Venn diagram, 81
verifiability, 50–51, 55–57, 61, 64, 100, 106–7, 110, 121
Virno, Paolo, 130n4
voice, 109, 129–30, 134, 138–39, 156–60, 170, 179
von Bertalanffy, Ludwig, 15
vote, 31, 54, 68–69, 136

Wales, Jimmy, 42, 49, 86, 93–94, 101, 105–6, 108, 121, 145, 150, 158, 160–79
ways of seeing, 42
Weber, Max, 9, 88–89, 93, 96–98, 110–12, 117, 119, 122, 124–25
Weber, Steven, 126, 133, 137–38
Web 2.0, 2, 3
Wegrzanowski, Tomasz, 162–65, 168–69
Wikietiquette, 54, 75
WikiLeaks, 3
Wikimedia Foundation, 10–12, 52, 59, 111, 148, 174
Wikiocracy, 124
Wikipedia, 3–13, 25, 27–28, 30–31, 38, 41–70, 74–87, 92–94, 98–127, 138, 177–81; art (*see* Wikipedia Art); article deletion, 8, 50–61; blackout, 10; fork (*see* forking; Spanish Wikipedia); gender gap, 11; images (*see* Muhammad); mediation procedures (*see* mediation [Wikipedia procedure]); organizing principles, 9; neoliberalism, 177–81; statement formation, 40; Web 2.0, 3, 4; working together (*see* collaboration; participation)
Wikipedia Art, 8, 50–61, 70, 74–75, 84–85, 103
wikis, 112
working model. *See* formal equivalence, working model of
World War II, 8, 15, 17–18
wrong, political, 77–78, 80, 82–83, 86, 136, 181

Xerox, 20–21, 23

Yahoo, 165

Lightning Source UK Ltd.
Milton Keynes UK
UKOW04f2120030215

245610UK00001B/3/P